Getting Under Our Skin

GETTING UNDER OUR SKIN

The Cultural and Social History of Vermin

Lisa T. Sarasohn

Johns Hopkins University Press
Baltimore

Johns Hopkins University Press
2715 North Charles Street
Baltimore, Maryland 21218-4363
www.press.jhu.edu

Library of Congress Cataloging-in-Publication Data

Names: Sarasohn, Lisa T., 1950– author.
Title: Getting under our skin : the cultural and social history of vermin /
Lisa T. Sarasohn.
Description: Baltimore : Johns Hopkins University Press, 2021. |
Includes bibliographical references and index.
Identifiers: LCCN 2020045435 | ISBN 9781421441382 (hardcover) |
ISBN 9781421441399 (ebook)
Subjects: LCSH: Pests—Social aspects—Great Britain and North America. |
Pests—Control—Economic aspects—Great Britain and North America. |
Pests—Psychological aspects. | Pests in literature. | Human-animal
relationships—Great Britain and North America.
Classification: LCC SB605.G7 S37 2021 | DDC 632/.960941—dc23
LC record available at https://lccn.loc.gov/2020045435

A catalog record for this book is available from the British Library.

*Special discounts are available for bulk purchases of this book. For more
information, please contact Special Sales at specialsales@jh.edu.*

CONTENTS

I started thinking about this subject when working on my last book, which was about the seventeenth-century natural philosopher Margaret Cavendish, Duchess of Newcastle. That remarkable woman did not fear approaching topics previously discussed only by men. She dared to challenge earlier interpreters of nature and members of the newly formed Royal Society. Her efforts, particularly her satire of the new experimental science and its champion, Robert Hooke, inspired me. In his 1665 *Micrographia*, Hooke included gigantic engravings of a flea and a louse. In Cavendish's 1665 romance *The New Blazing World*, these insects were reimagined as flea-men and lice-men. I wondered why Hooke had looked at fleas and lice and why Cavendish responded so viscerally to his images.

This book is the answer to those questions.

Once I started to think about vermin, I saw them everywhere: in every book I read—fiction and non-fiction—in every newspaper I opened, in every TV show I watched, even those not on PBS. Humankind seems engaged in an intimate, eternal dance with the creatures closest to our skins and our homes. I am a fan of *Star Trek* and Tolkien, and there vermin were again: the Borg, from whom resistance is futile; Shelob the spider, who all but consumed Frodo the hobbit. And then there were the fleas on my dogs and the rat in my garbage, making their own arguments. I thank, or at least acknowledge, all these sources for my perception of vermin.

Chapter 1 of the study first saw light as " 'That Venomous Nauseous Insect': Bed Bugs in Early Modern England," *Eighteenth-Century Studies* 46 (2013), 513–30, and "The Microscopist as Voyeur: Margaret Cavendish's Critique of Experimental Philosophy," in Sigrun Haude and Melinda S. Zook, eds., *Challenging Orthodoxies: The Social and Cultural Worlds of Early Modern Women:*

Essays Presented to Hilda L. Smith (Surrey, England: Ashgate Publishing, 2014), 77–100. I delivered parts of this book at the History of Science Society, the Renaissance Society of America, the Columbia History Society, the International Margaret Cavendish Society, and the Northwest Renaissance Society. I would like to thank all those who commented on both my publications and papers, particularly an anonymous participant at the 2010 Renaissance meeting who asked if I'd ever thought about lice and pornography. I hadn't then, but I certainly did afterwards. The book is much richer because of all these scholars and the anonymous readers for Johns Hopkins University Press.

I cannot thank my colleagues in the History Department of Oregon State University enough for their untiring interest and encouragement, especially Mary Jo Nye, Robert Nye, Paul Farber, Jonathan Katz, Anita Guerrini, Mike Osborne, and the late William Husband. Even after I retired, they provided input and criticism. Likewise, my fellow members of the International Margaret Cavendish Society, especially Brandie Siegfried and Sara Mendelsohn, have been sources of inspiration.

Librarians at the William Andrews Clark Memorial Library at the University of California, Los Angeles, the Huntington Library, the British Library, and the Folger Library have been unfailingly helpful. This book would not have been possible without the databases now available on the Internet: Early English Books Online, Eighteenth-Century Studies Online, Nineteenth-Century Studies Online, and Google Books. I can now sit at my desk and access a wealth of sources. Like many writers, I have found that this enables both narrowly-focused and broad studies of the past and present, a different scholarship from that produced by visits to the archives—whether or not an improvement, I leave to the readers.

At Johns Hopkins University Press, Matt McAdam has been a careful and insightful editor, and Will Krause and Julie McCarthy have been endlessly patient with my questions and the mystifying elements of Dropbox. Using artwork has been a brave new world for me, one not possible without help from them and my friend Julianne Bava. Ashleigh McKown's terrific copy-editing has saved me from many errors in transcriptions and citations. Hilary Jacqmin at Johns Hopkins was an invaluable help in working with the many illustrations in the book.

Lastly, I would like to acknowledge the editing skills of my husband, David Sarasohn, who has contributed to the substance of this book more than I can say. I appreciate the endless patience of my sons, Alex and Peter, who not only put up with a distracted mother but also provided Internet counsel. My grandson Theodore was gracious enough to allow me to work on this book when I should have been playing with him. I dedicate this book to all of them.

Getting Under Our Skin

Vermin in History

Vermin get a bad press.

The titles of recent books about vermin display a clear anti-biter bias: *Parasites: Tales of Humanity's Most Unwelcome Guests*; *Wicked Bugs: The Louse That Conquered Napoleon's Army and Other Diabolical Insects*; *Parasite Rex: Inside the Bizarre World of Nature's Most Dangerous Creatures*; *Pests in the City: Flies, Bedbugs, Cockroaches, and Rats*; *Rats: Observations on the History and Habitat of the City's Most Unwanted Inhabitants*. As legendary bug hunter Hans Zinsser declared in his 1935 classic *Rats, Lice and History*, emphasizing the deadly peril vermin present, "Swords and lances, arrows, machine guns, and even high explosives have had far less power over the fates of nations than the typhus louse, the plague flea, and the yellow-fever mosquito."[1]

But while vermin have undermined the power of mighty nations, they have burrowed even more deeply into the human psyche. Our response to vermin ranges from indifference to disgust, from laughter to fear, but it always reveals a society's view of what's acceptable or what's repugnant. Having insects on your body was once considered part of life, hardly meriting condemnation. Rats were always seen as repulsive, but more as an adept threat to agriculture than an indicator of filth and disease. When people no longer felt comfortable scratching in public or having rats in their pantries, when vermin—the creatures closest to us—were no longer seen as a universal burden but as a social stigma or existential threat, we had become modern.[2]

This book is the story of that transition.

Modernity, a slippery concept, is shown less by the advent of mass culture or advanced technology than by a growing disinclination to share our homes and bodies with pests, although what is considered a pest or parasite has changed over time. An exact date on this change can be as hard to catch as vermin themselves, but we can see it begin to bite around 1668, during a period historians call "early modern."[3] It then accelerated over the eighteenth century, with the rise of the middle class and with concern for hygiene, and burgeoned through the nineteenth century, often considered the start of "modern" times. The late nineteenth and twentieth centuries saw the identification of the germs carried by vermin, a medical advance, but also witnessed the ostracism and persecution of people considered verminous and disgusting.[4] Nothing could be more modern than the horror produced by the current bedbug incursions, which have been blamed on everything from the fall of the Berlin Wall to the Persian Gulf War to the spread of mosquito nets in Africa. Likewise, a notice that lice have been occupying the heads of preschoolers causes not only consternation but terror. Being modern has risks that our ancestors could not have conceived; perhaps when it comes to vermin, it was better to be a medieval peasant than a global, jet-setting citizen.

"Civilized" societies have viewed other peoples, both inside and outside their borders, as being a type of vermin—the ultimate other—socially, religiously, politically, and even sexually. This book traces who and what, at various times, were considered verminous, as well as the responses to the creatures themselves. Vermin became a prism through which to look at other people—generally downward. Bedbugs, fleas, lice, and rats are viewed as threats materially, morally, and metaphorically. Verminous people threaten the position of the upper classes, the power of the dominant group, and the comfort, health, and safety of the non-verminous. The disgust engendered by vermin—both animal and human—creates boundaries between "us" and "them," a necessary condition for contempt and conquest.[5]

Vermin highlight the vulnerability of the supposedly superior species. Vampiric, they feed on the blood and possessions of the targeted human. Bedbugs inhabit furniture and wall hangings, and attack while their victim sleeps. Lice occupy the hair and clothing of their hosts, branding them social pariahs. Fleas jump from person to person, carrying disease and dishonor. Rats bite babies and penetrate every human sanctuary, underscoring the precariousness of life and home. Vermin are emblematic of a world out of control, sometimes drawing horror, sometimes laughter—a rat dragging a slice of pizza down subway stairs is at once terrifying and funny. Truly, these creatures will survive any coming apocalypse or pandemic, outlasting mere humans. More visible than the microbes they carry, they become the vanguard of Pestilence, perhaps the most frightening of the Four Horsemen of the Apocalypse.

The identity of the verminous animals and humans has varied through time.[6] In the past, animals we would consider paragons of the natural world were considered vermin because they threatened the food supply or destroyed the wood used to heat or build homes. In *The Experienced Vermine Killer*, a 1680 book on how to get rid of verminous animals, the list includes foxes, moles, ants, snakes, caterpillars, worms, flies, wasps, bedbugs (then called wood lice), fleas, lice, otters, buzzards, rats, and mice, but it excludes rabbits, "though in former days accounted Vermine, are now preserved and much in esteem for the delicacy of their Flesh and Furs."[7]

In early modern Europe, vermin came to be viewed as a repulsive trait of the underclasses and foreigners. The powerful could easily redirect the weapons used against vermin—and the hatred and fear felt for them—toward the subjugated. Women could be ravaged, slaves flogged, prisoners condemned, Indians killed, foreigners conquered, Jews exterminated, all on the rationale that the verminous must be destroyed, or at least rendered powerless. In the eighteenth century, prostitutes were linked with pubic lice, slaves with fleas, bedbugs with the French, and native peoples with lice. The English were certain that the Scots and Irish—and later the peoples of the Continent—were uniquely infested, while European travelers reported nonwhite people living with and dining on lice, a clear indication of their inferiority and need to be conquered and colonized. As people came to see other races, nationalities, and classes as actually being vermin, genocide could be considered just a form of pest control.

The history of humanity's encounters with pests is not straightforward. In this century, amid widespread concern about the extinction of animal species, no one seems to think twice about killing fleas or poisoning rats. Their presence is greeted with a gurgle of dismay and a fast search for pest control services. In the past, however, the meanings tied to vermin were much more complex. Most of the time, people hardly loved vermin but saw them as representing both negative and positive qualities in persons and things. The changing meanings of vermin reflect changes in other attitudes. Fleas were admired for their dexterity. Rats had their own starring role in folklore, alternating between being the heroes and villains of stories. Lice, although never viewed as pleasant, could act as God's agents. Only bedbugs preserved their hideous reputation from the late seventeenth to the twenty-first century.

Throughout history, bugs and rodents have been a barometer of social norms, pressures to validate one's class identity, nationality, moral superiority, or masculinity. Vermin always signified something bigger than a bite. In the theology-driven Middle Ages, the suffering produced by pests provided dermatological confirmation of God's providential design. The sinful suffered because of lice, but some saints were considered especially holy because their lice- and flea-infested hair shirts provided mortification of the flesh, imitating the sufferings of Christ. In a famous story, the bystanders watching the martyrdom of Thomas Becket in the twelfth century alternated between weeping and laughter as multitudes of lice abandoned his cooling body. (Lice are particular about body temperature.) Centuries later, some moralists argued that God used vermin as a prod to cleanliness. In less lofty environments, village life was enlivened by the practice of mutual delousing, a way to keep fingers busy during nightly story times and to consolidate community ties. Vermin even crawled through upper-class thinking; noble ladies prized their lice combs, and poets envied fleas' freedom to venture beneath maidens' apparel.

Not that anyone—except perhaps saints and eccentrics—especially liked hosting vermin. Medieval and early modern housekeeping manuals are full of recipes for killing parasites, ranging from the dangerous (sulfur and turpentine) to the peculiar (roasted and ground dead cat). Nevertheless, even the most invasive parasites and rodents could be viewed positively. A whole volume of sixteenth-century poetry sang of a flea exploring a high-toned bosom. The metaphysical poet John Donne famously sought to seduce a young woman by arguing that a flea had already joined their blood together. Less poetically inclined young men longed for the liberty to lie around, living off the land and scratching. Explorers both marveled at and condemned foreign people who supposedly ate the vermin eating them. Travelers regaled a scandalized and enthralled Europe with tales of Indian chiefs fed crawling tidbits by their many strapping wives.

One did not have to go as far as the New World to find a more positive spin on pests, however. Countess of Eglinton Susanna Montgomery (1690–1780), considered one of the great beauties of her time, tamed rats that, on her command, would join her at dinner and "scamper off into their holes with great alacrity upon receiving the order to go. From the rats the Countess thought she obtained gratitude; an experience she believed to be very exceptional, and very rare in her dealings with the human species."[8]

But the countess was almost as unusual as the Indians who ate lice. Indifference toward bugs and rodents ended by the eighteenth century; most members of the upper classes cultivated a taste for baths and clean underwear along with a delicate aversion to the grubby. Increasingly, vermin were thought to destroy the integrity of the body or home. No longer was a bite an annoyance—now it became a source of corruption.[9] In order to protect the vulnerable, any method became allowable, hence the appearance of exterminators in the eighteenth century to eliminate the tiny destroyers of boundaries between vermin and humans.[10] During the same period, ratcatchers sought to establish their credibility and professionalism. One Thomas Swaine, "Ratcatcher to His Majesty's Royal Navy," published a guide to killing rats and mice, "those animals which are noxious to the community, describing their wiliness and sagacity, and the methods they take as well for procuring food and preserving themselves from danger."[11]

Increasingly, experts in the pest-killing arts supported their claims of expertise by citing their employment by the upper classes and by using pamphlets and advertisements in newspapers to get clients. Specialists emerged to expel bedbugs and lice from expensive beds and wigs. The very first exterminator, John Southall, advertised his services in 1730, promising to provide an elixir to kill bedbugs, "that Nauseous Venomous Insect." He died a wealthy man—his success often duplicated nowadays as vendors take advantage of the dread and humiliation brought by the recent bedbug onslaught. A Victorian tradesman billed himself as "the Queen's Ratcatcher, Rat and Mole-destroyer to Her Majesty." Another nineteenth-century exterminator advertised his business as "Tiffin and Sons: Bug-Destroyer to Her Majesty" and dressed in a kind of livery. Although he was never knighted, his grandfather did wear a sword and cocked hat when he went about on business.[12]

The destruction of vermin became a capitalist endeavor. The rising professionalism of exterminators reflected not only the concern for personal and domestic hygiene but also the reality of rapidly urbanizing localities that mixed populations to a disconcerting degree, at least for the upper classes. As the middle class defined itself by manners and purity, beggars, Gypsies, and shopkeepers were disdained as verminous.[13] Englishmen and -women and their colonial relatives tried to fortify the walls between classes by distancing themselves from the body and separating the homes of the better-off and the poor. In polite society, the butchering of animal bodies was removed to the slaughterhouse, and the human corpse was buried outside the town walls or enclosed in the dissecting chambers of hospitals and universities. Cleanliness became indispensable to social acceptability.[14] Servants proliferated in the effort to keep the home free of vermin, although the ladies of the house worried that their employees would bring vermin with them.[15] Ironically, newly installed sewers made it easier for rats to invade the bathrooms of the rich. Avenues were designed to act as barricades between the classes, but vermin were not so easily enclosed.[16] Nineteenth-century newspapers brimmed with advertisements for exterminators, and travelers complained incessantly about vermin in roadside inns and railroad terminals. The poor were even blamed for blocking progress—an early twentieth-century plan to electrify Duke Street in London faltered because of "disorderly boys, verminous women and tramps."[17]

One boundary that walls could not protect was the threshold of the flesh, which could be pierced by a bite or invaded by a smell.[18] The flesh becomes porous rather than protective when bitten, transforming the body into a victim of assault. Scratching an itch becomes a kind of participation in bodily invasion and disintegration—and hence a sign of lack of humanity. In order to protect the civilized person from such a fate, strict rules of etiquette developed in the eighteenth century, including the following precept, which an adolescent George Washington copied down: "Kill no Vermin as Fleas, lice, ticks, etc. in the sight of others."[19]

Thus bugs and rodents not only dissolve the borders of bodies but also penetrate the walls of social privilege, reincorporating newly risen members of the middle and upper classes with the multitudes they left behind. In a treatise on natural philosophy for young gentlemen and ladies, the female student remarks to her tutor, "but as to Bugs, Fleas, and Lice . . . Our very Bodies seem to be destined equally for their Habitations and their Food—is not this a most humiliating Consideration!"[20] And one royal rat killer, after recounting his adventures in the homes of gentlemen and shopkeepers, insisted, "Rats are everywhere about London, both in rich and poor places."[21]

Urbanization increased the risks of encountering the vermin ridden. Rats became the prime example of urban fauna. Henry Mayhew, a proto-social scientist who sought to describe all the inhabitants of the lower-class sections of Victorian London, spent many pages discussing rat catching and the newly popular sport of the rat pits, where patrons bet on how many rats specially bred dogs could kill. Likewise, flea circuses entertained the masses, whom their star performers bit. Perhaps this was a new way of controlling an upside-down world—one could at least laugh at its impositions.

Even those at the top of the social hierarchy had to contend with vermin that had the unsettling propensity to make them seem weak or absurd. Royalty and vermin may seem strange bedfellows, but serious and satirical literature abound with sagas of the high and mighty deflated by the small and biting. As creatures with an ecumenical appetite for the powerful as much as the impoverished, lice were a common tool for social commentary and political satire. From the time Moses used divine lice to make the pharaoh let his people go to the time George III found a louse on his dinner plate and a bedbug nibbled on his granddaughter, Princess Charlotte, vermin and rodents undermined authority, making it dependent on doctors, scientists, exterminators, and even slaves to conquer the tiny menaces.

From the beginning of the Scientific Revolution in the seventeenth century, researchers collected, studied, and catalogued insects in an effort to assert authority over their subjects and maintain the authority of their aristocratic patrons. Since most people thought viewing an insect under a microscope was an absurd waste of time, scientists had to defend their activities. Some argued that insects—even the ones as awful as lice and fleas—demonstrated the argument from design. God had created the very small as well as the very large. Even after eighteenth-century rationalism had undercut the purported role of the divine in nature, amateur natural historians, who were often clergymen, continued to reinforce their studies of insects and vermin with references to the creatures' creator.

But fleas and lice, roving from body to body and social class to social class, continued to defy trainers and observers (and exterminators), demonstrating the fragility of boundaries—and the ease of getting under people's skin. Often in history, the fear of vermin merged with fear of slaves and women, also weak creatures that could turn on their supposed betters—like insects before they were squashed or rats before they were trapped. In *Les Misérables*, convicts in a chain gang pick the vermin off their bodies and shoot them through straws at gaping bystanders. Weaponizing vermin could turn the weak into the strong, at least momentarily.

Entomology was called into the service of superiority. What had been a moral appraisal of the deficiencies and dangers presented by members of their own society—beggars, prisoners, and women—morphed into a condemnation of other societies and cultures, ultimately justifying colonization and racism. Skin color and entomology went hand in hand, with exterminators and experimenters arguing about whether dark or light skin was most prone to being bitten. But almost all agreed with the sentiments associated with Oliver Cromwell's policy toward the Irish and expressed again by US Col. John Chivington against warring Indians, "Kill and scalp all, big and little; nits make lice."[22] Even the smell of south Asian cooking elicited disgust. The odor of coriander—a spice abhorrent before the appearance of Indian restaurants—was considered foul, showing the extent that nauseating scents may be culturally specific.

By the nineteenth century, being verminous became a sign of not just social but also racial inferiority. Disgust turned into dehumanization. Entire nations, groups, and continents were fused with vermin and marked as inferior. In the twentieth century, the Chinese were associated with the spread of plague. The Chinese section of Honolulu was burnt down by white neighbors who feared infestation. In Texas, Hispanic maids were deloused before being allowed to cross the border.

Vermin contradicted the belief in progress so eagerly embraced by the dominant classes, particularly during the conflicts of the twentieth century. Soldiers during the World War I were often more—or at least more directly—worried about their "cooties" than about the bullets of the enemy, particularly those soldiers from the upper classes who had been brought up on the gospel of cleanliness. Bedbugs and lice and haunted the public housing projects that were supposed to solve the problem of housing the poor in the 1930s and proliferated in the bomb shelters of Britain during World War II.

Nineteenth- and twentieth-century discoveries that vermin carried typhus, bubonic plague, and other diseases only intensified the campaigns against those considered verminous. Internal enemies or foreigners were blamed for contaminating Western civilization. Lice became the common denominator in slaughter and genocide. When Reichsführer Heinrich Himmler identified lice with Jews, he was echoing the association of Jews with vermin that goes back to medieval times. It was no coincidence that the Nazis used the insecticide Zyklon B, developed during World War I to kill lice, in their gas chambers. Nevertheless, typhus carried by lice proliferated; when Allied soldiers liberated the concentration camps, it was too late for prisoners like Anne Frank, who had already contracted the disease. When Holocaust deniers argue that the concentration camps were simply delousing facilities—with no executions—they invoke an illusion and a link employed by the Nazis themselves.

Dichlorodiphenyltrichloroethane, commonly known as DDT, which was developed during World War II to destroy mosquitoes in the malaria-ridden swamps around Rome, gave humans a sweeping but temporary victory against vermin, annihilating various insects in the 1950s and 1960s. In "our next world war," exulted an editorial in *Popular Mechanics*, DDT would be the winning weapon against vermin, successfully fighting "a long and bitter battle to crush the creeping, wriggling, flying, burrowing billions whose numbers and depredations baffle human comprehension."[23] Echoing earlier racist tropes, the chief of the Chemical Warfare Service proclaimed, "the fundamental biological principles of poisoning Japanese, insects, rats, bacteria, and cancer are essentially the same."[24]

DDT was quickly commodified in the midcentury and was sold to housewives as a surefire aid to protect their children from vermin until it was banned in the United States in 1972 and Europe shortly thereafter. This was a triumph for environmentalists, but it was perhaps a less happy result for the many children infested with head lice ever after.

Whatever the results of the elimination of DDT—still used in developing countries to combat malaria—lice and fleas continue to proliferate in the West, as does the association between vermin and undesirables. In the 1950s, the Canadian province of Alberta proclaimed, "You Can't Ignore the Rat . . . Kill Him!" on posters that reflected the fear and hatred of border invasions by outsiders—rodent or otherwise. The cultural connotations of rats even infiltrated the work of midcentury American psychologists and scientists who extrapolated their lab findings about rats to human populations, particularly to African Americans. These analogies are no longer acceptable, but as cities gentrify in the twenty-first century, vermin are increasingly associated with immigrants and people experiencing homelessness. Nowadays, lice and bedbugs haunt homeless shelters, mass transit, and the facilities used to house immigrants, a source of concern for public health officials and a source of outrage for right-wing media.

The impact of the verminous on the human psyche—and body—echoes throughout modern English usage. Our vermin-infested past gave us the ubiquitous adjective "lousy," used to label everything that's bad or disgusting. Likewise, the skin-crawling ambience of medieval inns endures in the negative reviews of so many hotels as "flea-bags," hopefully metaphorically. When Sherlock Holmes warns that the world is not ready for the story of "the Giant Rat of Sumatra" or James Cagney (however apocryphally) addresses another mobster as "you dirty rat," they express, in varying vernacular, the fear of the loathsome other. And when the cartoon character Yosemite Sam addresses his rodent foe Bugs Bunny as "varmint," before trying to shoot him, he mirrors centuries of anti-vermin, anti-other sentiments.

These days, guns may be the only weapon not used against vermin. Currently, everything from mayonnaise to lice combs are used to combat the enemy. In spaces where the battle is clearly being lost, flea bombs, spreading chemicals seemingly every bit as scary as DDT, are set off as a last resort. In the luxury high-rises of Manhattan, specially trained dogs sniff out bedbugs, and an iPhone app uses GPS to detect where they lurk. The public sector has also intervened in the bedbug wars, inspiring the US Environmental Protection Agency to hold two National Bed Bug Summits. The New York City Rat Information Portal provides a city map that shows areas to avoid if you don't like rats—although, since rats infest the subways, you may not be able to go anywhere anyway.

Bedbugs, now resistant to many insecticides, have reappeared even at the highest levels of society. What is the point of being in the top one percent if you have to fear exposure as a helpless victim of a tiny intruder? Even the *New York Times* seemed unhinged when reporting the assaults of bedbugs on the wealthy: "Bed bugs on Park Avenue? Ask the horrified matron who recently found her duplex teeming with the blood-sucking beasts."[25] Businesses have responded with innovations to curb the infestations, advertised in gendered language whereby males attack the fiends and females protect the home from them. Politicians have rallied to the cause of bedbug annihilation, and they become enraged when they have been likened to this enemy. Lawyers have made careers out of representing clients suing the hotels where they claim to have been bitten. Representatives of the Group of Seven countries may think twice about staying at the Trump National Doral Miami as the president suggested; a guest successfully sued the Trump organization after being bitten there, a suit that was settled before the 2016 election.[26]

In many cases, the war against vermin is an overreaction, reflecting centuries of fear and revulsion rather than actual danger. But rationality has never been the hallmark of vermin–human relations. Giant bugs and rat-like creatures star as humanity's enemies in so many science fiction movies because the modern mind sees them as the ultimate other. On the big screen, these creatures suck, drain, feed on, and bloody the body—or in the case of the most famous (and chillingly insect-like) *Alien*, grow like a parasite inside the human body itself, consuming vulnerable flesh like its smaller inspirations.

One reason the fear of bugs and rodents is so pervasive in the twenty-first century is that it is so commercially rewarding. Science fiction movies make money, as do exterminators and the makers of bug repellent. Cartoonists and writers profitably use these creatures as emblems of human foolishness and human vulnerability. The Internet is awash with up close and personal images of bedbugs and roaches. In the seventeenth century, when bugs were first magnified many times by the microscope, viewers responded with horror—long before the terrors set off by nature documentaries.

Pervasive images of vermin in popular culture reflect their continuing cultural power. Since 1960, *Mad* magazine has carried the Cold War parody *Spy vs. Spy*, where two rat-like figures kill each other in fanciful ways, only to reappear, vermin-like, the following month. (Rats transcend even mutually assured destruction.) More profoundly, the award-winning graphic novel *Maus*, by Art Spiegelman, casts Nazis as cats and Jews as mice. In this illustrated world of the Nazi death camps, meaning is expressed and horror conveyed through the medium of vermin.

But we also use cartoon vermin to try to reassure ourselves. The Disney-fication of mice and rats—Mickey and Minnie Mouse, Remy the rat in *Ratatouille*—reimagines these vermin to be not carnivorous but cuddly. Mickey and Minnie carry on like any young couple—the mouse even has a pet dog—while the movie rat, far from biting people, feeds them gourmet meals. Cartoons are modern fables, and in these fables, vermin—domesticated and humanized—cease to threaten humankind.

In *Thinking with Animals*, Lorraine Daston and Gregg Mitman argue that the modern preoccupation with other creatures arises both from our efforts to create a kind of community with them and from the urge to use animals "to symbolize, dramatize and illuminate aspects of their own experiences and fantasies."[27] Any observation—verbal or visual—of a bug, a louse, or a rat conveys a world of meaning about the attitudes, prejudices, and jokes of any particular place or time. Vermin's very closeness—biting us, inhabiting us, eluding our control—makes them the animals holding up the sharpest mirror to ourselves and the universe we share with them—whether we want to or not.

When I began working on this book, I had no idea that pestilence, whether caused by fleas and rats carrying the bubonic plague or typhus spread by lice, would have any immediate resonance for me or my readers. These diseases seemed something that truly belonged to history, in the modern vocabulary of something that no longer applies. But I have been reminded that nothing is really past, especially when it comes to the human response to calamity. Now, in the age of the coronavirus pandemic, some believers are wondering whether COVID-19 is the first rider of the Four Horsemen of the Apocalypse, released by the Lamb of God to announce the beginning of the end of days.

Less theological but equally disconcerting, Mike Pompeo, Donald Trump's secretary of state, and some people in the conservative media have called the coronavirus the Wuhan or Chinese Virus, implicitly linking the disease with the oriental other set on destroying Western Civilization (as we know it!). Trump himself has referred to COVID-19 as a "foreign disease" or the China disease. The fact that coronavirus jumped from animals—perhaps bats—to humans has intensified the idea that the disease and those who have it are bestial and should be separated from society. Social distancing has been introduced as a palliative, but we know from history that isolation measures, even medically necessary ones, often can lead to social antipathy. The ghettos of Europe did not cause the Holocaust, but they did make it easier to divide people into us and them. When the president of the United States calls the governor of Washington State—where the coronavirus outbreak was the most serious in the country at one time—"a snake," we are well on our way to a society like the one described by the Florentine humanist Giovanni Boccaccio in 1348 when the bubonic plague hit the city: "Caring for nobody but themselves, many men and women forsook the city, their own homes, their parents, and friends, and fled. They seemed to believe that the wrath of God, punishing men with his plague, would fall on none save those inside the city. Or they believed that the ending of all things had come."[28]

This book reflects many of the current trends of historical scholarship, but also the experience of anyone scratching a bite. Historians have started to study the history of the body and the history of animals, along with the ways the two came together. Perhaps because of our increasingly materialistic culture, the matter of history has changed. We continue to want to know about the legendary dead white men, but we also want to know how ordinary people lived—and how they scratched. At the borders of the flesh, the external world and the internal body interact when vermin penetrate the skin. The home itself is an extended body, occupied by transgressive rodents. People get nervous when boundaries become threatened, and that anxiety manifests itself in efforts to guard and control the invaders and illegals who threatened the integrity of the individual and his property. Some of these efforts seem strange, some seem familiar, and some seem funny, but all of them guide us through humanity's verminous history.

People often ask how I came to be interested in such a loathsome topic as vermin. The short answer is that I was impressed when I saw the images of a flea and a louse, the late seventeenth century's first fruits of the microscope, in Robert Hooke's *Micrographia*, and I wondered why he chose to picture these particular animals in the first book devoted to showing the newly revealed microworld. The long answer is more complex. It combines my own fear of bugs (although not the ones I write about here; I really hate spiders) and my training as a historian of science and ideas. My previous book was about Margaret Cavendish, the first woman to write about natural philosophy, science's label in the seventeenth century. She wrote *The New Blazing World*, a romance in which Hooke's fleas and other animals are metamorphosed into beast-men organized into scientific societies by her heroine. Her goal was to satirize, but she also showed how significant animals were in the imagination of past times.

Once aware of insect and rodent references, they become ubiquitous, and not only in the literature and documents of the past. They inhabit poems, novels, newspapers, and even cookbooks. The current bedbug mania is only the most recent example of how pervasive and significant vermin are. Almost everyone has a lice story—the memory and horror inspired by the insect endure long after the lice have crawled off. Anybody who has met a rat exploring their garbage, as I have, knows what true terror feels like. And when environmentalists explain that the rat and the roach will inherit the earth, we can only agree that their chances look good.

The term *vermin* includes many species of animals. Sometimes it is narrowed to parasitic insects, and sometimes it is expanded to include mice, rats, and other mammals (including some humans). The multiplicity of vermin required picking and choosing. I decided to concentrate on the insects that live in our bodies, clothes, and beds—bedbugs, lice, fleas. Less attention was paid to mites, ticks, and mosquitoes in history, so they also get less notice here. As it's hard to keep rats out of any structure, they have nosed their way into the book.

"That Nauseous Venomous Insect"

Bedbugs in Early Modern Britain

Bug. A stinking insect bred in old household stuff.
—Samuel Johnson, *A Dictionary of the English Language* (1768)

In all the centuries preceding the eighteenth, being bitten in bed was a fact of life. When Samuel Pepys mentions vermin-infested beds in his diary, he seems amused rather than abused. On one trip, he and a companion got "Up, finding our beds good, but lousy; which made us merry." During another journey with Dr. Timothy Clerke, Pepys described their nightly experience: "We lay very well and merrily; in the morning, concluding him to be of the eldest blood and house of the Clerkes, because that all the fleas came to him and not to me."[1]

Pepys was clearly unsurprised and unfazed by his nightly encounters with bugs; the good-natured lack of distress about the insects feasting on him and his friends is evident. So is the joke about social class; the parasites prefer Dr. Clerke because of his presumably bluer blood. But jokes about class and vermin became increasingly problematic in the eighteenth century, as the upper classes became more conscious that bitten bodies had social and cultural meanings beyond the necessity to scratch. The bedbug, in particular, elicited horror as a newly recognized pest that breached the bodies and bedrooms of its eighteenth-century hosts. In the search for modernity, the bedbug may be the canary in the coal mine, indicating the changing attitudes toward body and environment that characterize modern society.[2] The reaction to this nighttime terror reveals much about the mentality, prejudices, assumptions, and aspirations of society in the eighteenth century.

A hundred years after Pepys deemed night-biting vermin a joke, the playwright and novelist Oliver Goldsmith viewed bedbugs with horror: "The Bug is another of those nauseous insects that intrude upon the retreats of mankind; and often banish that sleep, which even sorrow and anxiety permitted to approach. This, to many men, is, of all other insects, the most troublesome and obnoxious." Cried Goldsmith, sounding almost paranoid, "It is generally vain to destroy one only, as there are hundreds more to revenge their companion's fate; so the person who thus is subject to be bitten, remains the whole night like a sentinel upon duty, rather watching the approach of fresh invaders, than inviting the pleasing approaches of sleep."[3] Rather than a joke, bedbugs were a mortal enemy.

Insect parasites were legion in early modern England, but the presence— and effect—of verminous insects on the body, and on the cultures that bodies constituted, continues to be overlooked. Historians have neglected the role that these predators played in early modern times except as disease vectors, perhaps because we have largely (although increasingly temporarily) excised their presence from modern Western societies. Verminous insects were part of the lived experience of earlier humans as well as an extension of and intrusion on their bodies. While we cannot ever know how people in the past experienced insect bites on a phenomenological level, we can know how they understood their uninvited lodgers.[4] In recent decades, the embodied human being has become a focus of anthropological, philosophical, feminist, and historical literature, much of it "confusing and contradictory," as the historian and MacArthur Fellow Caroline Walker Bynum has written.[5] But all these approaches share the belief that our perception of the body is culturally constructed. The body does not simply exist as a biological entity but is both the product and producer of culture.[6] The body is not just the embodiment of individual autonomy and self-consciousness but a mediator between oneself and the other; indeed, no human being ever lived—or thought—without a body.[7]

The most threatening of all attackers on the body, at least in terms of eighteenth-century reactions, were bedbugs. Being bitten by a bedbug was repulsive, humiliating, and nauseating—even its smell elicited disgust. The intense reaction to bedbugs reflected the growth of a global economy and the consciousness of race in a society becoming more urban, middle class, and xenophobic. Once the bedbug entered popular parlance as "the bug," it could be conflated with insect exotica and commodified as a threat one paid to be rid of. The battle against bedbugs invoked an emerging expertise culture and the creation of a service industry—exterminators—to employ against the menace. These conquistadores of the insect world were a tool used by the upper and rising classes to defend themselves against a threat thought to arise from less ordered peoples and places. Exterminators sought to establish their authority by associating themselves with the scientific and societal institutions of the eighteenth century.

Thus the first scientific examination of the bedbug—the 1730 *A Treatise of Buggs* by the exterminator John Southall (available for one shilling at the Oxford Arms in Warwick Lane)—was dedicated to the president of the Royal Society, Hans Sloane. It included the following *apologia*: "I have ever since my return from America made their [the bedbugs'] destruction my Profession . . . [and] determin'd by all means possible to try if I could discover and find out as much of their Nature, Feeding and Breeding, as might be conducive to my being better able to destroy them. And 'tho in attempting it I must own I had a View at private Gain, as well as the Publick Good, yet I hope my Design will appear laudable, and the Event answer both Ends."[8] The private and the public good, science and service, knowledge and conquest all are complicit in Southall's claim that during his sojourn in America he had discovered an elixir to kill bugs and safeguard England from a dangerous threat.

In the rich range of vermin that plagued early modern bodies, bedbugs were perceived as a newly identified and particularly dangerous menace. Before 1500, bedbugs were often confused with other insects, at least in terms of nomenclature. In England, by the sixteenth and seventeenth centuries, there were many references in published works to wood lice or wall lice, which clearly refer to the bedbug.[9] The naturalist Thomas Muffet, for example, in his 1658 *The Theater of Insects*, describes wall lice by using the Latin *cimex*, an insect that "seeks after living creatures that are asleep" and that "omits a most abominable stink" when it is touched.[10] Likewise, John Ray, a naturalist and member of the Royal Society, described the insect he encountered in Italy:

> Cimei as the Italians call them, as the French Punaise. We English them Chinches or Wall-lice, which are very noisome and troublesome by their bitings in the night time, raising a great heat and redness in the skin. . . . This insect if it be crushed or bruised emits a most horrid and loathsome scent, so that those that are bitten by them are often in a doubt whether it be better to endure the trouble of their bitings, or kill them and suffer their most odious and abominable stink. We have these insects in some places of England, but not many, neither are they troublesome to us.[11]

It was only at the end of the seventeenth century that the insect formerly known as the wood louse, cimex, punaise, or chinch was rechristened the "bug" or "bugge" and suddenly began to bother the English as much as they did foreigners. The new name gave bedbugs a new identity and a new capacity to terrify, like the apparitions with whom they share a similar name, bugbears. A bugbear, according to the *Oxford English Dictionary* (OED), is "An object of dread, especially of needless dread; an imaginary terror . . . [it is] an annoyance, bane, thorn in the flesh." While the OED cannot trace the etymological link between bug and bugbear, the two terms parallel each other in their connotations of terror and vulnerability.[12] One might argue that bedbugs are bugbears made flesh—the actual thorn in the flesh.

The Odor of Bedbugs

Living bodies are not just composed of flesh; the senses include smell, a border that is even more difficult to police than skin. Eighteenth-century Englishmen and -women were particularly offended by the odor of bedbugs. What did it mean that bedbugs smelled bad to the noses of discerning (and not-so-discerning) early modern noses? It might be that lice and fleas were so common that people became desensitized to their presence, but bedbugs appalled in an olfactory sense. They did not merely bite—they smelled, and the pungency of this early modern odor in the nose of the eighteenth-century English needs to be explained. The perceived evil odor of bedbugs becomes a cultural marker of not just the noses but also the minds of early modern men and women, particularly those who wanted to distinguish themselves from the lower classes, foreigners, and other outsiders.[13]

The bedbug, or the *Cimex lectularius*, is part of the order of hempitera or winged insects, although it is not winged. It belongs to the family Cimicidae, which is found in both in Europe and the Americas and feeds mostly on bats. Humans and bedbugs probably first began their association in the caves of the Paleolithic Age in the warmer climates of the Middle East and Mediterranean, eventually spreading to northern Europe during the central Middle Ages. In modern times, bedbugs are found everywhere, but they need a warm place to thrive—hence bedbugs and well-built dwellings go hand in hand. They are hard to see before they have fed, when they are pale brown and very thin, but after becoming engorged with blood, they grow larger (to about one-quarter of an inch) and turn a brick red color. Most modern entomologists believe that bedbugs, unlike their fellow vermin the flea and the louse, carry no diseases.[14]

The nauseating smell of bedbugs is the most common referent to this creature in early modern texts. The disgust at their smell did not dissipate through the eighteenth century. Thus Goldsmith, in resonant and vivid prose, complains, "Nor are these insects less disagreeable from their nauseous stench, than their unceasing appetites. When they begin to crawl, the whole bed is infected with the smell; but if they are accidently killed, then it is insupportable."[15] The naturalist and instrument maker George Adams agreed: "These insects are as disagreeable from their nauseous stench, as their unceasing appetites."[16] The French physician and pharmacist Louis Lémery, whose encyclopedic work on food and health was translated into English in 1745, noted that the coriander plant "has an unpleasant Smell, like that of Buggs, and that is the Reason that 'tis neither us'd in Physic, nor Food."[17]

Smells, particularly odors perceived as disgusting, are culturally specific. One culture's smell of sanctity (the odor of saints) becomes another culture's sign of putrefaction.[18] Early modern Englishmen were not renowned for their sensitivity of smell, routinely putting up with the stench of open sewers and rotting offal, but the perception of the bedbug's loathsome odor reflects the almost hysterical reaction this kind of vermin elicited. Fleas and lice, other vermin that inhabited human bodies, were terrible but sometimes considered funny, and they were sometimes used as vehicles for satire and religious edification. Pharaoh in Egypt suffered a plague of lice, and George III had a louse fall on his dinner plate, but bedbugs had no such royal and divine provenance.[19] They were simply odiferous and disgusting.

The seventeenth-century writer Thomas Tryon (1634–1702), perhaps the world's first health guru, believed in the curative power of fresh air, and he linked bedbugs and smell so closely together that he assumed that the insects were produced by the odors themselves. He wrote, "From the pernicious Smells and putrified Vapours that do proceed from old Beds, are generated the Vermin called Bugs, (of which, neither the Ancients, nor the Modern Writers of this Age, have taken any notice) according to the Degrees of Uncleanness, Nature of the Excrements, and the Closeness of the Places where Beds stand." And he continued by saying, "Stinking Scents and Vapours which do proceed from the Bodies and Nature of Men and Women, and the mixing or incorporating of these Vapours with moist and sulphurous Airs: For where there is no Heat nor Humidity, there can begin no Putrefaction."[20]

The spontaneous generation of insects from matter was an old and increasingly discredited theory by the time Tryon was writing in 1682.[21] But Tryon's argument relies on another perception of disease that became only stronger in the eighteenth and nineteenth centuries. Miasma, or noxious vapors, caused not only plague, according to medieval and early modern sources, but also other forms of illness.[22] Thus Tryon could argue that "From the same Substance or Matter whence Bugs are bred, is also occasioned the Generation of many nasty diseases."[23] Smelly bugs were both disgusting and lethal.

Bedbugs and the Rise of the Middle Class

The sensitivity to the nauseating smell of bedbugs may also reflect the rise of the aspiring nose. The reputed pungency of their odor displayed the growing sensibilities of the urban middle class and perhaps exemplified the arrival of this perennially rising class, or at least its olfactory delicacy, by the early eighteenth century.[24] It also paralleled the growing disapproval of the odiferous and filthy lower classes who now smelled bad to their betters.[25]

The desire to be clean is as culturally determined as the sensitivity to smell. Before the eighteenth century, clean hands, faces, and visible linen indicated cleanliness—but clean undergarments and bedding were optional. Rapidly, dirt or nastiness became associated with the lower and working classes. An early eighteenth-century satirical text that ridiculed bankrupt shopkeepers charged, "they would sit, some Raving, some Muttering, some Laughing, & others Gaming, till drunk and drowsie, they reel home to their dirty Rooms, Sheetless Beds, & and spaul'd Garrets to feed the Flees, as well as worse Vermin, till the next Morning at which time they return again like a dog to his Vomit, or a Sow to her wallowing in the Mire."[26] The worse vermin are undoubtedly bedbugs.

The association of vermin with shopkeepers indicates that bedbugs and other parasitical insects had particular ties to different social classes. Lice and fleas, expected by all classes in the earlier centuries, were presumably pickier in the eighteenth century, particularly as cleanliness became a desideratum and a marker of the upper classes. By 1682, Tryon was instructing "Cleaness in Houses, especially in Beds, is a great preserver of Health."[27] The physician Richard Mead (1673–1754) wrote in 1720, "It is of more Consequence to be observed, that as Nastiness is a great Source of Infection, so Cleanliness is the greatest Preservative: Which is the true Reason, why the Poor are the most obnoxious to Disasters of this Kind."[28]

But bedbugs did not respect social privilege and middle-class morality. They invaded the homes of the affluent, whom they ate just as readily as they consumed the poor. Their presence produced not only itchiness, but also humiliation. Hence the increasingly alarmed response to the presence of bedbugs becomes even more understandable.

Prosperous Englishmen and -women worried that servants would introduce bedbugs into their homes, with disastrous results. *Read's Weekly Journal* reported in 1760 that a maid had set fire to her master's house while trying to burn bugs out of her bed.[29] The same year, another maid had inadvertently killed a porter whom her master had commanded she provide with drink: "Upon enquiry it was found that the maid had given him a glass of liquid that was bought a day or two before, in order to destroy buggs, instead of the brandy."[30] The fear of bedbugs and their social consequences extended to the claim that the lower classes were weaponizing the insects (even if not, as in the sad instance noted above, killing people). The assault of bugs on the body had now metamorphosed into the attack of one person on another. Another porter was accused in 1733 of purposefully seeding a bathhouse with bugs. A newspaper described the trial of the accused:

> At the Westminster Sessions of the Peace an Indictment was preferr'd against one Robert Speare, a Porter, at a Bagnio near Covent-Garden, for bringing a Number of live Buggs in a Bottle to the Bagnio House of Mr. Bates in Pall-Mall, and throwing them secretly under his Beds, with Intent to destroy his Furniture, but the Jury return'd it Ignoramus.[31]

We remain as ignorant as the jury as to the circumstances of this alleged crime. Was Mr. Speare trying to get back at his employer, or was he acting on his employer's orders to destroy the business and reputation of a competing establishment? The wariness of intentional infestation was also demonstrated in 1733 in Ireland—supposedly previously free of bedbugs—when a correspondent warned a member of the Right Honorable Dublin Society, "We hear there was another Person whose Head had a very Mischievous Turn, and who valued himself for bringing over a little Box with Buggs, which (to prevent our Sleepiness) were turned to breed in our Beds."[32] By 1736, the bedbugs had spread to Scotland, where another hapless woman burnt down her house while trying to rid her bed of them.[33] And back in Ireland in 1749, bugs in beds produced more legal action when a woman sued a tradesman for selling her an infested bed—the Lord Mayor ordered her money refunded and the bed burnt.[34]

A bed was an expensive item in the eighteenth century. But this possession, which signified middle-class ascendency, paradoxically made owners more vulnerable to its eponymous bugs. Nothing certified status more than the middling classes' material possessions, and prominent among them were elaborately carved beds that welcomed bedbugs as much as human beings. Since bedrooms were used for entertaining as well as sleeping, the smell of bedbugs might be particularly disturbing to a newly gentrified tradesman or professional man—or his wife—attempting to establish his position.[35] The cabinetmaker William Cauty advised those in the unfortunate possession of a bug-infested bed, "if you can spare your bedsteads for some weeks or months, commit them naked to the yard or out-house, which will be of singular use, excessive heat or excessive cold being the absolute cure of all vermin."[36]

Cauty directed his advice to "gentlemen of fortune" who perhaps slept on expensive feather beds, increasingly popular among the upper classes during the seventeenth and eighteenth centuries. Some thought these beds were also guilty of bedbug propagation. Tryon indicted feather beds for both their moral and verminous qualities: "old stinking Feather-beds," he tells us, "which possibly stunk before ever they were lain on ... do certainly contain an unclean putrified Matter, that hath a near affinity with the Nature of Bugs; and therefore feather-beds are more apt to breed them, than Wooll or Flocks." Feathers are "unclean, fulsome Excrements, of a hot strong Quality."[37]

Foreign Bedbugs

Tryon's argument that bedbugs were the result of smells, sweat, or feathers indicates how Englishmen sought an explanation for the appearance of this apparently new and revolting insect. One theory of their origin, endorsed by John Southall, is that bedbugs accompanied the foreign timber imported to rebuild London after the Great Fire of 1666. Southall says that he questioned "as many learned, curious, and antient Men as I possibly could, concerning them," and they endorsed the following explanation,

> That soon after the Fire of London, in some of the new-built Houses they were
> observ'd to appear, and were never noted to have been seen in the old, tho'
> they were then so few, as to be little taken notice of; yet as they were only seen
> in Firr-Timber, 'twas conjectur'd they were then first brought to England in
> them; of which most of the new Houses were partly built, instead of the good
> Oak destroy'd in the old.[38]

Likewise, William Cauty, some fifty years later, blamed foreign influences for the introduction of bedbugs, but he was even more specific than Southall. "It is hard to ascertain," he wrote, "how this species [of insects] was first introduced into England: probably they came with the French refugees, who fled from the persecution of Louis le Grand." His answer to the problem is a return to good old English wood: "Now to make a bedstead, sopha, or chair, so as no vermin can exist in them; take the best English Oak, or any other good sound wood, for your bedstead frames" and "anoint every part with spirit of turpentine, mercury and spirit of salt."[39]

The foreign origin attributed to bedbugs reflected the growing sense of national superiority that accompanied the rise of British power and riches in the eighteenth century. But bugs and other kinds of vermin threatened to undermine British naval dominance, as much as the French and the Spanish. Infestations of ships—and the sailors who manned them—were a constant worry of the Admiralty. In a sense, the nautical body was as subject to attack as the physical body. One report on the state of the warships Fourgeaux and Monarchy stated that one-third of their bedding had to be destroyed, another third sold in public auction, and the last third "well-cleaned, scour'd and air'd" because they had infected "Men laid in them with a great Degree of the Itch."[40]

Vermin even figured in the propaganda war that led to the War of Jenkins' Ear between the English and the Spanish in 1739. In a letter published in the anti-Court *Royal London Evening Post* in March 1738, an Englishman named Thomas Bruch claimed to have been held in a Spanish prison near Cadiz where he was forced to work like a slave and to eat "som rotten Benes full of Buggs, a Sampell of wich I shall bring with me, if ever I shold be so happy as to see England." The newspaper particularly lamented the failure of the English merchants at Cadiz to help the poor prisoner: "The unhappy writer complains too of the Neglect and Hardheartedness of the English Merchants at Cadiz, to their countrymen now devouring by Vermin, and stifling with Filth in Dungeons just under their Noses." It concluded, "If any Briton can read this without a heavy Heart and a wet Eye . . . you will fully discover whether you are really a Briton, or a Spanionized Englishman."[41]

The association of vermin with foreigners allowed the English to disconnect themselves from the entomological other, implicitly emphasizing the distinction between British and continental bodies or states. Goldsmith captures this patriotism in his description of the onslaught of bugs.

> These are part of the inconveniences that result from the persecution of these odious insects: but happily for Great Britain, they multiply less in these islands, than in any part of the continent. In France and Italy the beds, particularly in their inns, swarm with them; and every piece of furniture seems to afford them a retreat. They grow larger also with them than with us, and bite with more cruel appetite.[42]

Both the Cambridge don John Martyn and the encyclopedic naturalist Francis Fitzgerald incorporated Goldsmith's patriotic entomology into their works.[43] William Cauty also emphasized the dangers that Englishmen faced from bedbugs on the continent, making travel across the Channel sound almost as perilous as a journey up the Amazon. He tells an anecdote about "A gentleman travelling once thro' Italy, lay in as fine and clean apartment as could be," but still "he was often devoured, in a manner, by them; he averred, they dropped from the ceiling [*sic*], perpendicularly upon him, and were often gone before he could get up, returning to their holes in the corners of the room, or cornices, where there might be some rends in the plaister walls."[44] In this case, not even the absence of wooden bedsteads can save the Englishman from the ravages of voracious Italian bedbugs, which lurk in the borders and crevices of rooms waiting until sleep has made his body vulnerable to their attacks. The traveler could have used the advice of Mariana Starke, an Englishwoman traveling through Italy in 1797: "It being necessary when you travel on the continent, to carry your own sheets, pillows, and blankets. I would advise the doubling them up daily of a convenient size; and then placing them in your carriage, by way of cushions, making a leather sheet around the envelope." Upon arriving at an inn, Starke suggested, "Four or five drops of essential oil of lavender distributed about a bed, will drive away either bugs or fleas for the night."[45]

Fighting Bedbugs at Home and Abroad

Mrs. Starke's recommendation of lavender to combat bedbugs is one of the pleasanter remedies proposed against the nauseating insects, including William Cauty's recipe of a mixture of mercury, salt, and turpentine. As a kind of sympathetic magic, using strong smells to vanquish other strong smells, household manuals often included turpentine and sulfur in their recipes to combat insects. *The Universal Family Book*, in 1703, recommended mixing rue with storax and turpentine, and advised, "shut the Room, or Chamber as close as possible, and sprinkle these on Chafing-dishes of burning Charcoal, and go out to avoid the Scent your self, and shut the Door after you, and the strong Scent will kill them."[46] The physician Boyle Godfrey tried using turpentine and arsenic, as well as rue, wormwood, and other chemicals, to kill bedbugs, but he had the most success with sulfur mixed with "Oyl of Vitriol." He heated it up, closed up his bedroom, and lit the mixture on a charcoal stove. The resulting "prodigiously strong Funck" was "such as will kill all Creatures in the Universe."[47]

Less benign, but probably also less effective, were a series of recipes for killing bugs described in an English edition of the French *Dictionaire* [*sic*] *oeconomique*, by Noel Chomel, reissued many times during the eighteenth century. Following the strong smell philosophy, it recommends "the Smoke of Cow-Dung," rotting cucumber, and ox gall mixed with vinegar to combat bedbugs, but it is the last item that may have caused Godfrey to remark, "As to all the pretended Remediesor Cures against these Vermin offered the Publick daily, I perswade myself they are nothing but Frauds."[48] Instead, Chomel suggested, "Kill a Cat by stifling her, without drawing any blood from her, and having taken off her Skin, and her Guts out, roast her upon a Spit, without larding or basting; and keep the Stuff that drops from her, which you are to mix with the Yolks of Eggs and Oil of Spike in equal parts; let them be well incorporated in a Mortar till brought to the Consistence of an Ointment."[49] Once you have the ointment, sufferers should rub the mixture onto the parts of the furniture frequented by bugs. Chomel doesn't advise the sufferer what to do if these extreme measures are ineffective, but he could perhaps have borrowed Cauty's advice about his own remedies not working: buy a new house.[50]

Self-help often did not do the job, and so in the eighteenth century the quest to rid beds of bugs gave rise to a whole new profession: exterminators. Like other emerging professions during this period, bug killers attempted to establish their authority through testimonials from the upper classes and attacks on one another's credibility. They claimed to be experts at their jobs and described their methods as scientific and established by experimentation.[51] Experts were often denounced as quacks by their competitors, accused of conning a credulous public into accepting their panaceas.[52] In the fluid society of early modern England, nascent exterminators negotiated their status, trying to differentiate themselves from craftsmen and mountebanks, and thus establish successful businesses and increased social standing.

What we might call the battle of the bug killers started early in the eighteenth century and can be traced through competing advertisements in the ubiquitous newspapers of the period. All of these ads claimed the approbation of upper-class clients who usually were not named; presumably, they would have been less than pleased to have their bedbug woes exposed to the public. A certain John Williams advertised, "I have for several Years, with Success made it my Practice to destroy those nauseous Vermin called Buggs. At a reasonable Price; and those noble Persons who have employ'd me, are very well satisfied with my Skill and Performance." The advantage of his service, he claimed, was that it was "done without the least Damage of Hurt either to Bed, Bedding, or Furniture, be the same be ever so good; and what is used is without any offensive Smell."[53] What could be better than a cheap service whose success is attested by nobility—and that avoids the loathsome smells and dangers to good furniture of other remedies? But another pest killer, George Bridges, did him one better. While repeating the claim that he will not hurt the furniture or leave noxious odors, Bridges added that "he has cured upwards of ten thousand Beds . . . and be the Buggs ever so numerous or intolerable, he will warrant . . . to destroy them so as never to return again, or breed any more in the same Beds during Life. No Cure no Money." Bridges claimed to have a "Secret" that would destroy bedbugs for at least ten years. To authenticate his claims, he noted, "a Character of his Performance may be had in any Neighbourhood all over the Great Metropolis, London, from People of the best Repute and Quality."[54]

George Bridges hoped that giving a money-back guarantee and testimonials for his services would protect him from being called a "Puff, Cheat, or Imposter." London seems to have had more than enough bug killers seeking to profit from the distaste for bedbugs, and some among them were clearly quacks. One presumably legitimate exterminator took out an advertisement against a rival who had set up business near him in order to "save his Reputation, and his Customers not be mistaken."[55] The best way to do this was to up the ante and social rank of one's character witnesses and identify them by name. William Cauty claimed that Lord Egmont, late the Lord Admiral, would endorse his services.[56] And in 1785, according to a satirical essay in the *British Magazine and Review,* among "the numberless absurdities" that characterized London were the many tradesmen claiming "to be thought the servants of his Majesty." Among the many different professionals claiming the king as their patron were many "bug-destroyers." The author of the article agreed that "the palace of St. James's is an antiquated building, and that, as it is for the most part adorned with wainscot, a mouse or a bug may at times have the presumption to intrude on the royal premises; yet I think one person employed in each of the those honourable departments would, by proper exertion, be adequate to the task of destroying all these nauseous vermin."[57]

At the same time that bug hunters were citing their ties with the upper classes, they also seemed to imply a racial susceptibility to bedbug bites. William Cauty began his 1772 treatise *Natura, Philosophia, & Ars in Concordia* with an endorsement of the mechanical philosophy and described his bug-killing prescription as a "method." He claimed to be unlike "Those who pretend to cure bedsteads and chambers from vermin" who are like "the quacks, who heal the outside of a sore, but leave the canker-worm inside." He claimed to be a scientist of bugs, giving weight to his connection of race with bedbugs. Cauty argued,

> There are many skins which they will not touch: thick and dark skins are not their favourite soil: a stranger who has a fair skin, stands a chance of being bitten by them: several persons have in one night's time been so stung by them, as to be left next morning, like one with the small pox, and in an high fever besides.[58]

Fair skin, he asserted, was more liable to bedbug bites than dark skin, and the bites that white people endured were worse than those inflicted on the darker races. Complexion suddenly became a component of the discourse on bedbugs, especially when the English confronted the native people they conquered and enslaved in Africa and the New World.

According to the physician and botanist Patrick Browne, Jamaica teemed with bedbugs.[59] And from Jamaica the most famous of the early exterminators brought his expertise and experience in dealing with bedbugs, "That Nauseous Venomous Insect."[60] John Southall, on undetermined business in Jamaica, suffered greatly from bedbug bites, avoided only when a freed African slave gave him the recipe for a lethal insecticide.[61] He published *A Treatise of Buggs* in 1730, and a second edition was issued in 1793. Another exterminator, John Cook, issued a challenge to Southall's remedy in 1768, discounting it because "like too many more narrow-minded men, [he] had not the generosity enough, by discovering the secret, to render the public proper proof thereof." Like many purveyors of elixirs—and many early capitalists—Southall preferred to keep his formula a secret. But worse than lack of public mindedness, Cook charged that Southall had learnt the composition of the "liquor from a certain Negro, with which Stygian water" Southall had claimed "he can destroy those loathsome insects, and all their eggs also."[62]

In Cook's account, the race of Southall's source was used to undermine his credibility, but the original discoverer of the *non pareil liquor* clearly believed his account of learning the formula from an elderly slave (who was more than seventy years old) in Jamaica would help sales. The Englishman was tapping into one of the beliefs of colonizing conquerors of the New World: it and its inhabitants could be coerced to reveal secrets that would aid the Old World. Southall's account shows him bribing the African with food and drink to learn his secret. Southall himself, in a reversal of the dominance and submission typical of European and native interactions, then certifies the black man's authority: "'tis my Opinion, he had attain'd to a greater knowledge of the physical Use of the Vegetables of that Country, than any illiterate Person ever had done before him."[63] Southall had found an "African Magi" whose intimate knowledge of plants and animals could be appropriated for his own benefit.[64]

On his return to England in 1726, Southall opened his bug-killing business for profit and, according to this nascent entrepreneur, the aid of the public. He gained the sponsorship first of John Woodward, Gresham Professor of Physic at Cambridge, and after Woodward's death of Sir Hans Sloane, president of the Royal Society. Sloane was a good choice as patron. He had published his own work on Jamaican natural history after visiting the country as a young man, and he collected specimens from the New World during his entire career.[65] Southall was validating his work with the big guns of eighteenth-century science, apparently with their full support. Sloane introduced his protégé to the Royal Society, where on January 8, 1729, the exterminator read his *Treatise of Buggs* and received their "Approbation," which he was careful to cite at the beginning of the published work in 1730.[66]

The treatise itself includes a long description of the morphology of the bedbug. In writing what was in actuality a long advertisement for his exterminating business, the entrepreneur became an entomologist. According to Southall, his scientific treatment of bedbugs was motivated by the persistence of the vermin he pursued that managed to survive even the formula provided by the Jamaican. He "determin'd by all means possible to try if I could discover and find out as much of their Nature, Feeding, and Breeding, as might be conducive to my being better able to destroy them."[67] In good Baconian fashion, this tradesman decided to study his product—clearly, the empirical as well as the imperial imperative had spread widely by the first part of the eighteenth century. He invoked science to separate his work from the scams of other purveyors of bug remedies; an advertisement for his business proclaimed him "the first and only Person that ever found out the Nature of Buggs."[68]

In the treatise, Southall describes how he had purchased a microscope and observed the development of bedbugs from nit to maturity. An illustration of this progression serves as the frontispiece of the pamphlet, demonstrating the increasing importance given to graphic imagery at this time. The picture as witness is another source of authority in science, and these engravings have the power to make the observer scratch.

In the text, Southall explains how the six-legged bug sucks "our Blood, their most delicious Food," through a sting with which "they penetrate and

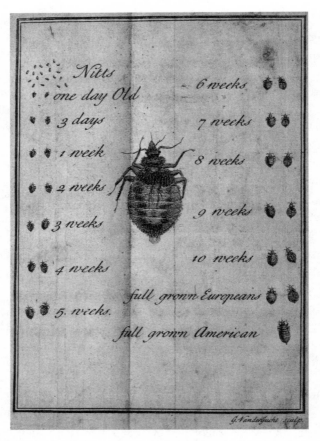

Frontispiece of John Southall's *A Treatise of Buggs*, 1730. Project Gutenberg, https://www.gutenberg.org/files/49564/49564-h/49564-h.htm

wound our Skins." Thus they actually suck rather than bite us. Bedbugs change color as they age, going from milk white to brown in seven weeks, becoming full grown by eleven weeks. Southall distinguishes captured bugs from wild bugs, and the larger American bugs from European bugs, although "when they spawn and breed here, the Young degenerate, and are of the European Size."[69] The wild bedbug is especially vicious, at least toward his fellow insects: "Wild Buggs are watchful and cunning, and tho' timorous of us, yet in fight one with another, are very fierce; I having often seen some (that I brought up from a day old, always inur'd to Light and Company) fight as eagerly as Dogs or Cocks, and sometimes one or both have died on the spot."[70]

Tame bedbugs are more malleable and more social. After enclosing several pairs in bottles, Southall observed them spawning "about fifty at a time," of which about forty lived to three weeks. Since they spawn in March, May, July, and September, he explained, "'tis apparent to a Demonstration, that from every Pair that lives out the Season, about two hundred Eggs or Nits are produc'd; and that out of them, one hundred and sixty, or one hundred and seventy, come to Life and Perfection."[71] While live bugs cannot live through the winter, their nits, being "inanimate," can persevere through the cold months and become more active by heat or fire. Their lethargy during the winter makes it "the best Season for their total Destruction." All those who had attempted to eradicate bedbugs before, Southall argues, including "many People of Sense and Learning, as well as the Vulgar and Illiterate," failed because they did not understand that quiescent nits survive in the wood and wainscoting of homes during the cold, only to move to the furniture and beds when spring comes.[72] Therefore Southall only guaranteed his process to work during the winter, because "if cleared out of Spawning-time, there is a certainty, as there is then no Nits, that their Offspring cannot plague you thereafter."[73] This then was Southall's version on "if there are no nits, there can be no spawn."

Southall was eager to dispel all of the myths surrounding bedbugs, including their supposed preference for one kind of person over another. "In reality," he maintains, "they bite every Human Body that comes in their way; and this I will undertake plainly to demonstrate by Reason." Arguing by analogy to the infection caused by cuts or wounds, the aspiring experimentalist explained that bedbug bites become inflamed only when a person has "an ill Habit of Body" rather than "a good Habit of Body," and in fact the inflammation resulting from a bite indicates when someone lacks a "right order of Blood." Thus the mystery is solved, "That where two Persons lie in one Bed, one shall be apparently bit, the other not."[74] It seems that Samuel Pepys would no longer be able to ascribe the attack on his friend to "blue blood" but rather to bad blood or perhaps to bad character, which returns us again to the moral imperative of cleanliness.

John Southall was a practical as well as enterprising individual. He included good advice for those seeking to avoid bedbug infestations—much of it similar to that given during the twenty-first-century bedbug plague. He urged people to carefully inspect their furniture and luggage for marks left by bedbug excrement, and also to check the boxes and bags that servants bring from other places of employment. In the latter bit of advice, social class and bedbugs are linked. The baskets of washerwomen are particularly dangerous and can introduce bugs into clean linen. Southall advises that furniture, especially bedsteads, should be plain and made of hardwoods like oak, rather than softwoods like fir and pine. Old and used furniture should be avoided, and carpenters and upholsterers should not recycle used materials. If, however, all these prophylactic measures fail, Southall was willing and able to sell his "Nonpareil Liquor" for only "2 s. sufficient for a common Bed, with plain Directions how to use it effectively," with prices going up according to the furniture or rooms treated.[75] If the customer did not want to do the work himself, Southall continued, "You may have it expeditiously done by me or my Servants, and your Beds, or such Part as is necessary, taken down and put up again in full as good, if not better Order, than they were before, and alter'd, (if I see Opportunity or Occasion) and made to draw out, on my usual easy Terms."[76]

Southall died a wealthy man, and his wife continued his business for some time after his death.[77] *Treatise of Buggs* was republished in 1793, with the business sections excised and a recipe for the destruction of bedbugs added by "A Physician." The formula includes mercury chloride (otherwise known as the surefire cure for syphilis) and sal ammoniac boiled together with various herbs, turpentine, and wine dregs. Southall claimed, "If a live bug is but touched with a drop of this mixture, you will see it die immediately," and moreover he claimed it safe to rub on furniture.[78] It is doubtful that this product worked any better than any previous formula. As we will see in chapter 2, the exterminator Mr. Tiffin was still doing a land-office business in the nineteenth century, and bedbugs plagued the trenches in World War I and the bomb shelters in World War II. They were finally vanquished by DDT after the war but have since become resistant to that lethal concoction. The foremost twentieth-century expert on the history of parasitical insects, J. R. Busvine, presciently warned in 1976, "Yet there is still some danger to a recrudescence. In many hot countries, bugs have become virtually immune to DDT, because of resistance. If some of these resistant bugs are introduced, in the luggage of some immigrant, they could spread and provide a serious control problem."[79]

Bedbugs and their foreign carriers are still the outsiders who threaten the security and sleep of the modern—and indeed postmodern—world. The eighteenth-century response to the onslaught of this kind of vermin is instructive. Bedbugs caused more than a bite. Their presence signaled that the newly powerful classes, and the furniture that signified their rising status, were still vulnerable to the tiniest kind of threat. The material goods that the rising middle classes wanted to possess could turn on them and thus reveal the precariousness of their claims to higher status. Their servants could bring the bugs in with the wash or even use the vermin as weapons against their employers. The urban dweller, happily sleeping in his new bed, was literally overwhelmed by the smell of bedbugs, which he fruitlessly sought to stifle and contain.

Bedbugs invaded England, and the English blamed the attack on foreign imports; the products of the New World had unintended consequences, as colonization and travel were not simply a matter of appropriating and profiting from other places. Ambivalence undermined confidence in dealing with the natives of other lands, where native knowledge might provide the key to defeating an enemy now within. The superiority of the English to other races and skin types became problematic when bedbug bites affected fair people more seriously than the darker peoples. The Royal Navy worried that their ships and sailors would become unusable because of the prevalence of vermin—bedbugs became a national security threat.

European science and medicine had limited success in dealing with the bug, allowing a whole new profession of experts to arise to deal with these problems. Exterminators relied on the upper classes to certify their work, and they battled among themselves about who was legitimate and who was a quack. They offered their clients a new service and a new commodity: an exotic treatment to protect Englishmen from a menace from abroad. The insecticides they developed went from the benign to the dangerous in their ingredients and applications. Not surprisingly, by the middle of the eighteenth century, newspapers were advising the wise consumer to build in stucco rather than wood, bypassing the threat of the bugs or their remedies.[80] Another technological fix was proposed by an English surgeon who suggested using iron bedsteads after observing them in Italy. Should English hospitals decide to employ them, they would hopefully provide comfort "to so many thousands of miserable wretches, that are tormented sometimes even to death, by these nauseous vermin."[81]

Insects had always held the lowest place on the great chain of animal being, but this particular insect was perceived to be particularly noisome and nauseating.[82] In the post-Cartesian world of the eighteenth century, animals could be dismissed or used, but vermin impinged on the distinction of human and animal that Descartes had made a cultural maxim in the seventeenth century. Their insistent agency as they breached the boundary of the body, and their odor that penetrated the nose, inverted the dominance of humans over the natural world accepted as natural and inalterable by early modern Englishmen. Bedbugs seemed to destroy all barriers—between man and animal, between the interior and exterior of the body, the gap between the bug-ridden masses and their hygienic (and thus moral) superiors. They put all pretentions to superiority at risk and therefore became the parasite most feared and attacked.

The response to bedbugs was unique and problematizes the human–animal relationship during a period of rapid change in social mores. Bedbugs, and their smells and scars, were the physical and metaphorical incarnation of eighteenth-century fears about race and class, globalism and commerce, medicine and quackery. The bedbug—in Southall's words, "That nauseating, venomous Vermin"—crawled into the consciousness of the modern world and exposed all its vulnerabilities. In the dark of the night, the bedbug leveled the eighteenth-century English gentleman's elaborately constructed claim to preeminence over the entomological other and demonstrated how exposed he was to many kinds of threats.

CHAPTER TWO

Bedbugs Creeping through
Modern Times

Crazy Bernie [Sanders], he's crazy as a bedbug, but, you know, he doesn't
quit.

—Donald Trump, June 17, 2016

In his entomological assessment of Bernie Sanders, presidential candidate
Donald Trump was reflecting an age-old legend. Using an expression going
back to the early nineteenth century, Trump mocked Sanders's somewhat
unusual appearance, emphasizing his supposed lunacy as well as his dogged-
ness. Like bedbugs, which scurry every which way when a light shines on
them—and are really, really hard to get rid of—Sanders plagued Hillary Clin-
ton and may have cost her hours of sleep.

Bedbugs are back. After nearly disappearing in the late twentieth century
owing to the use of DDT, they have emerged as the most potent insect threat
to the sanity of human beings, if not to their health. Infesting homes and
possessions, bedbugs can make sane people crazy. This can be literally true;
at the 2011 meeting of the American Psychiatric Society, a scientist studying the psychiatric effects of bedbugs reported that infestations can cause
"a wide variety of affective, anxiety, and psychotic spectrum illnesses causing
significant impairment, including suicidality and psychiatric hospitaliza-
tion."[1] Not surprisingly, the news media is rife with stories about the lengths
people have gone to in attempting to rid themselves of the beasts, including
but not limited to smoke, fire, dousing their belongings with alcohol (do not
light a cigarette when doing this, as a teenager in Detroit learned the hard
way), and using pesticides in a quantity that would kill an elephant—if not
always a bedbug.[2]

Bedbugs seem preternaturally intelligent. The urge to anthropomorphize
these predators is almost impossible to avoid. They are "sneaky" and "me-
thodical," according to a leading trade magazine of the pest control industry.[3]
The patterns of their bites are characterized as breakfast, lunch, and dinner,
and when their habitats are disturbed, they "walk out the door, down the hall
and through the door into the next apartment."[4] Sufferers feel like they are
being tortured; a contributor to the *Washington Post* agonized, "I feel like

someone broke into my house. And started living there. And had sex in my bed."[5] Bedbugs not only bite, they also violate the most intimate places we live, subjecting us to home invasion or a kind of insect rape, even if we spend thousands of dollars to defend ourselves.

The response to the current bedbug plague echoes the hysterical reactions of eighteenth-century Londoners. In the early nineteenth century, the bedbug was viewed, according to the naturalist and cleric William Bingley (1774–1823), as "a nauseous and troublesome inhabitant of most houses in large towns," sucking the blood of sleeping people.[6] Jane Carlyle (1801–66), the wife of the writer and social critic Thomas Carlyle (1795–1881), wrote him in 1852 that "The horror of these bugs quite maddened me for many days."[7] At a time when entomology was becoming established as a separate scientific discipline, and even flies and termites could elicit a good word from naturalists, bedbugs were the most hated insects in nature.[8]

Throughout the nineteenth century, scientists, entrepreneurs, entomologists, and home economists in both England and the United States continued the war against bedbugs. Reporting on an 1833 exhibition at the American Institute of the City of New York, an association of inventors, a newspaper expressed: "Mr. Thomas Miller deserves most of his country for a Steam Engine, which he calls a 'Bedbug Exterminator,' and which, for aught we see, will 'use up' these remorseless cannibals pretty efficiently. If he can take his invention into Virginia and rename it the 'Chinese Destroyer,' he is a made man. We rejoice in common with the afflicted in both hemispheres upon the prospects of a mastery over these villainous insects."[9]

Why this invention should be particularly useful in Virginia and why it should be renamed a "Chinese Destroyer" is somewhat mysterious. By the early nineteenth century, trade with China was increasing, but there was little consensus on how to view the Chinese. Some Europeans and Americans saw the Chinese as masters of invention, including gunpowder, paper, and printing. Chinese porcelain goods were especially prized. But others saw China as a decadent and filthy country. The author of this piece seems to be sympathetic to the idea that a bedbug exterminator could help people in both the West and the East, but there is also the connotation that the Chinese may have been exporting the insect as well as cups and vases. Is he equating the Chinese with bedbugs and therefore suggesting that both the bug and the Chinese be destroyed? One business in the nineteenth and early twentieth centuries called itself the Oriental Exterminating Company, perhaps again linking the onslaught of the pest with the Far East. Once more, visions of vermin became visions of other people.

The exotic origin of bedbugs, and their association with the other—whether foreign, political, or racial—extends the associations Englishmen made between bedbugs and the darker races, whether Black people, Italians, or French, and with events like the Great Fire of London. At the same time, the steam engine of the nineteenth century actually increased the possibility of bedbug infestations, as railroads and hotels swarmed with the bugs. Entrepreneurs eagerly sought to provide travelers with products to defend themselves, with companies flourishing in urban centers in England and America.[10]

Urbanization, cleanliness, and class continued to shape bedbug condemnations into the twenty-first century. In 2010, an Internet news site proclaimed, "Bed bugs are swiftly becoming the bane of humanity, plaguing homes, office, movie theaters and malls alike."[11] Bedbugs are particularly prevalent in large cities, with Baltimore, Washington, DC, Chicago, New York, Chicago, Los Angeles and Columbus, Ohio, topping the infested list in 2018.[12] Bedbugs are currently the most vivid symbol of urban blight, except perhaps for rats. Between 2004 and 2010, New York City complaints about bedbugs soared from 537 to 10,985.[13] Since 2007, the *New York Times* has published dozens of articles on the pest, including many on the mystery of their current surge and how their reappearance has been linked to various ethnic groups. Unsurprisingly, different cultures blame different groups for creatures: the English blame Eastern European workers, Eastern Europeans point a finger at the Roma, Americans blame immigrants, and everywhere the poor and alien are accused of bringing the pest into the homes of the affluent.[14]

Just as with nineteenth-century railroads, travel is a risk in the twenty-first century. Travelers are warned to inspect, and even take apart, their beds in hotels, and to never leave their bags on the floor. Those so unfortunate as to bring bedbugs back from vacation are instructed to either freeze or heat their possessions.[15] Apparently, Thomas Miller was on the right track with his invention of the Bedbug Exterminator steam engine.

The reaction to recent bedbug infestations also has cultural resonances. Londoners in the eighteenth and nineteenth centuries smelled an enemy that could threaten their social aspirations. In the twenty-first century, it is the danger of social opprobrium that terrifies most; according to the *New York Times*, sufferers "feel ashamed, even traumatized, to have these invisible vampires living in their homes . . . they feared the condemning glares of neighbors or the shunning of co-workers."[16] Unlike the past, when it was not surprising to have royal palaces infested with bedbugs, and exterminators publicly naming their noble clients, secrecy is mandatory in the current crisis. Landlords and upscale hotels seek desperately to keep their bedbug troubles quiet, forcing the bitten to sue for damages or insisting on nondisclosure clauses in sales contracts. Class seems to be even more important now than in the stratified society of eighteenth-century London—bedbugs should not be biting the urban upper classes. The *New York Times* reported, "Beyond the bites and the itching, the bother and the expense, victims of the nation's most recent plague are finding that an invisible scourge awaits them in the form of bedbug stigma. Friends begin to keep their distance. Invitations are rescinded."[17]

And, as in the eighteenth century, people have turned to experts and science to save them from the bedbug menace. Scientists now know a great deal about bedbugs, including the fact that they do not carry diseases, at least not in the microbial form.[18] Bedbugs can crawl through the leavings of other insects or animals and promote illness that way, but they are dispersal units, not instigators themselves. Their medical effects are psychological and not physical, although about one-third of people bitten by bedbugs may have allergic reactions and huge welts—one dermatologist refers to them as "juicy bites"—that disappear in seven to ten days.[19] In rare cases, numerous bites can cause anemia, although generally this occurs in people already immunocompromised.[20] Others have no reaction all. The bedbug anesthetizes the skin before sucking up blood, so most people don't even know they are being bitten. The signs of bedbugs are mostly indirect, like drops of blood or trails of feces on one's pillowcase. The bugs themselves are difficult to recognize. Entomologists and exterminators often assure the panic stricken that the bugs terrorizing them are only beetles or stink bugs. The fears of one anxious client reflect how powerfully the tiny bedbug feeds on our imagination—"Please," she pleads, "let it be scabies," not bedbugs![21]

The comedian Amy Schumer captures the many resonances of modern bedbugs in a passage from her memoir, *The Girl with the Lower Back Tattoo*: "Not only did I move into the worst part of Queens, but I also got bed bugs, the 9/11 of bugs, which provide both a logistical and existential nightmare when they come into your life. They are nearly impossible to get rid of—so I had to subject my poor elderly stuffed animals to a scary ride in a high-heat dryer. And everyone I knew was quietly reevaluating their friendship with me."[22]

The Science of Bedbugs

The naturalists William Kirby (1759–1850) and William Spence (1783–1860), in their *History of Entomology*, which first appeared in 1826, agreed that is not a good thing to find bedbugs co-sleeping in one's bed. They argued that bedbugs had been introduced into England by trade from abroad:

> Commerce, with many good things, has also introduced amongst us many great evils, of which noxious insects form no small part; and one of her worst presents were doubtless the disgusting animals now before us. . . . But however horrible bugs may have been in the estimation of some, or nauseating in that of others, many of the good people of London seem to regard them with the greatest apathy, and take very little pains to get rid of them; not generally, however, it is to be hoped, to such an extent as the predecessor of a correspondent in Nicholson's *Journal*, who found his house so dreadfully infested by them, that it resembled the Banian hospital at Surat, all his endeavours to destroy them being at first in vain.[23]

Like every describer of bedbugs from the seventeenth century on, Kirby and Spence agree that their subject is nauseating. They also anticipate the disgust of future naturalists, who will view the indifference of Londoners as a sign of their laziness or apathy, perhaps a less-than-Christian attitude toward the poor in the devout nineteenth century. Kirby was one of the many nineteenth-century cleric-naturalists of the time who studied nature as revealing God's design. Like the Swedish entomologist Carl van Linné (1707–78)—better known as Linnaeus, the naturalist who first developed the two-part Latin names for classifying creatures—Kirby saw a divine harmony in nature, not necessarily including bedbugs.[24] Kirby greatly admired Linnaeus's major work, the *Systema Naturae*, published in nine volumes between 1735 and 1768. The 1802 English translation by William Turton explained that a bedbug was "A troublesome and nauseous inhabitant of most Cities: crawling about in the night-time to suck the blood of such as are asleep, and hiding itself by day in the most retired holes and crevices."[25]

In 1819, a dictionary reprinted this definition verbatim, demonstrating how quickly natural history penetrated popular literature in a time of widespread fascination with nature. The dictionary further depicted the bedbug: "In sucking it is a perfect glutton, never ceasing unless it is completely gorged and can hold no more."[26] Judgment on the bedbug, and on those whose homes

Cimex lectularius, an adult bedbug in the process of eating. Centers for Disease Control and Prevention, Public Health Image Library #9820

harbored it, was dispensed by many, including the Harvard entomologist and botanist Thaddeus William Harris (1795–1856), who maintained that the insect inherited its name from the "bugbear" of earlier times, "objects of terror and disgust by night."[27] Modern science has revealed the bedbug in all its glory (and goriness). Bedbugs are particularly unappetizing after consuming their meal of blood. In one feeding, they can suck up enough blood to last them several days. If their host (or, less graciously, nutrition source) vanishes, the bedbug can hibernate without feeding for more than a year. As the eighteenth-century bedbug aficionado John Southall knew, bedbugs can survive a cold winter and reappear with the sunshine. In fact, it takes a dryer set on the highest possible heat for thirty minutes, or a home with a temperature of 118 degrees for seventy minutes, to definitively kill bedbugs—and as the University of Minnesota Extension Service cheerfully informs, that doesn't prevent bedbugs from re-infesting a location at a later date.[28]

Consequently, the hunt for an effective method of getting rid of bedbugs is constant. The exterminators' old solutions, from roasted dead cats to mercury, faded only slowly from folk memory. In the nineteenth century, mercury,

turpentine, gasoline, and arsenic were used to kill bedbugs, and the twentieth century saw pyrethrum, hydrogen cyanide, and several other chemicals
added to the arsenal.[29] Unfortunately, most of these treatments killed only
mature bedbugs, leaving their offspring to flourish another day. It was DDT—
described as "the atomic bomb of the insect world" by the US Centers for
Disease Control—that really annihilated the disturbers of sleep. Bedbugs—
which inhabited almost one-third of homes in London before World War II
and happily roomed with patrons of railway inns throughout the late nineteenth and twentieth centuries, not to mention barracks, movie theaters,
brothels, and any common meeting places—were vanquished without the
bother of squishing them.[30]

But as early as 1976, the pioneering entomologist J. R. Busvine warned
that the bedbug was increasingly resistant to pesticides and poised for a
comeback.[31] The resurgence was bolstered by the banning of DDT in most
developed countries in the 1970s, and by 2008, bedbugs had become common companions in households and businesses. Entomologists suddenly find
themselves sought after, and exterminators are viewed almost as saviors.
While modern bug killers do not roast cats to take on bedbugs, they do use
specially trained dogs to hunt them out. Not surprisingly, dog trainers have
been accused of falsely identifying bedbug infestations when none exists—
bedbug treatments are lucrative for the treaters.[32] There is now a smartphone
app that uses GPS data to inform subscribers where the bedbugs lurk, and a
special vacuum equipped with ultraviolet light that kills the nits and nymphs
but not the mature bedbug. As Southall taught, "If there be no Nits, there can
be no Spawn."[33]

And scientists are busily seeking other methods. Recent attempts involve
using bedbug pheromones to lure the insects into bug-killing desiccants, described in a study in the *Journal of Medical Entomology*.[34] Some entomologists claim that pheromones can also craze adult male bugs, driving them
kill immature bugs; perhaps this is what John Southall observed in the behavior of wild bedbugs. Other methods, including using otherwise banned
insecticides, can be found online at the site of the First (April 2009) and Second (February 2011) National Bed Bug Summits.[35] Authority in science is now
linked to government support; public service is the responsibility of the public sector. The First Bed Bug Summit was instituted by the US Environmental
Protection Agency (EPA) and set up a Federal Bed Bug Workgroup, including the EPA, the US Department of Housing and Urban Development, and
the Departments of Agriculture, Commerce, and Defense.[36] Just as the Royal

Navy worried about bedbugs' threats to English shipping and defense, the US federal government hopes to keep it from undermining national security.

Class, Race, and Bedbugs in the Nineteenth and Early Twentieth Centuries

In the eighteenth and nineteenth centuries, the public often depended on the good will of specialists like John Cook and William Cauty to battle the bedbug menace. The couple in Isaac Cruikshank's cartoon on the facing page, "Summer Amusement: Bugg Hunting," does not seem particularly terrified or disgusted by their pursuit of bedbugs, but they certainly seem like members of the dirty lower classes. They wear slovenly and ragged clothes while rats lurk on their windowsill. On the wall of their bedroom is a print of an advertisement by T. Tiffin, Bug Destroyer to Her Majesty, perhaps whose aid the couple will solicit when their own efforts prove fruitless.

In the mid-nineteenth century, the scion of a family of exterminators, Mr. Tiffin, spoke to the journalist and proto-social scientist Henry Mayhew.[37] Mr. Tiffin dated his family business back to 1695, started by an ancestor who was a ladies' stay maker, probably familiar with bedbug bites. This Mr. Tiffin advertised his business as "Tiffin & Son, Bug-Destroyers to Her Majesty" during the "Illumination for the Peace" following the Peace of Amiens in 1801. The "majesty" mentioned was Princess Charlotte, in line for the throne if she outlived her grandfather George III and her father George IV. (She didn't.) The unhappy princess had called for his services to destroy a bug that had been biting her during the night. When he found it, she exclaimed, "Oh, the nasty thing! That's what tormented me last night; don't let him escape." Mr. Tiffin reported to Mayhew, "I think he looked all the better for having tasted royal blood."[38] (Mr. Tiffin, like many other Englishmen, seems to have had a jaundiced view of the couth and refinement of their Hanoverian rulers.)

The Tiffins claimed to work only for affluent and prestigious clients. The exterminator insisted, "I have noblemen's names, the first in England, on my books." His family's service to the upper classes was so much part of its self-fashioning that his father, according to Tiffin, used "to go to work killing bugs at his customers' houses with a sword by his side and a cocked-hat and bag-wig on his head—in fact, dressed up like a regular dandy." After much experience, Tiffin argued, "I've never noticed that a different kind of skin makes any difference in being bitten." All are prey to bedbugs, which he claimed to eradicate by method rather than the supposed remedies of other bug killers:

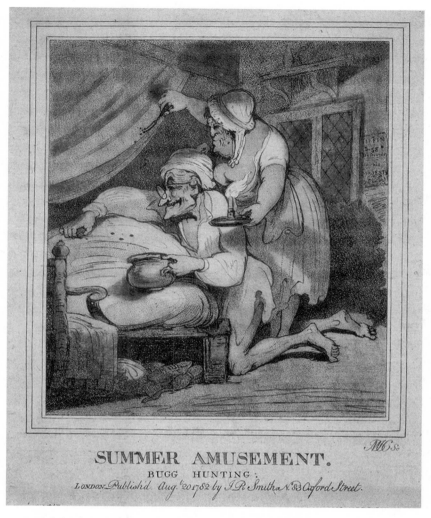

Isaac Cruikshank, *Summer Amusement: Bugg Hunting*, aquatint, 1782. Library of Congress Prints and Photographs Division, control number 00652100

"I may," he said, "call it a scientific treating of the bugs rather than wholesale murder. We don't care about the thousands, it's the last bug we look for, whilst your carpenters and upholsterers leave as many behind them, perhaps, as they manage to catch."[39]

Implicitly, Tiffin was claiming the same refinement that he said characterized his customers. He asserted supremacy owing to sharing the scientific propensities of many gentlemen in the eighteenth and nineteenth centuries who took to natural history with passion and exactitude. He did not bludgeon his bugs; he eliminated them not with a hammer but a rapier, perhaps like his sword-bearing father. He was a focused assassin, not a mass murderer like some lower-class exterminators. And like other collectors of natural objects, Tiffin collected unusual specimens: "I have plenty of bug-skins, which I keep by me as curiosities, of all sizes and colors. . . . There are white bugs—albinos you may call 'em—freaks of nature like."[40]

In Tiffin's quest for status, even the bedbugs he pursued were special. It is worth noting that he preserved their skins rather than their whole bodies. The close association of bugs and the skin they penetrate may have operated in the talismanic sense, making his collection not merely a source of research but of natural magic. Like the exterminators who used strong smells to destroy bugs, Tiffin is using an aspect of the other-enemy—in this case its skin—to control it. His live bugs, in addition, were worthy opponents: they "colonize anywhere they can, though they're very high-minded and prefer lofty places." In this case, both the insects and their hunter share a taste for glory, if not for blood. Paradoxically, however, and contrary to his claim that bedbugs had no preference in skin type, Tiffin mentioned that "The finest and fattest bugs I ever saw were those I found in a black man's bed. He was the favourite servant of an Indian general."[41]

Tiffin's hobby reflects the natural history pursuits of many of the more professional entomologists, like the naturalists Kirby and Spencer. The nineteenth century was awash in amateur naturalists who eagerly collected examples of flora and fauna, both at home and in the territories the British colonized and conquered. Like Southall, they were seeking to control the natural world for both profit and status. Collecting the odd and exotic was a theme of emerging capitalism, a way to display and accumulate wealth, a pastime of empire builders and international financiers.[42] The Tiffins were so successful that their business still exists today, although it has been renamed—perhaps predictably—Ecotiffin, providing a multitude of building care services.

Paralleling the profits of bedbug pursuers, bedbugs themselves grew fat on the blood of town and city dwellers. The children's author Beatrix Potter (1866–1943) described her experience of smelling bedbugs while vacationing in the seaside town of Torquay in 1883: "I sniffed my bedroom on arrival, and for a few hours felt a certain grim satisfaction where my forebodings were maintained, but it is possible to have too much Natural History in a bed."[43]

Since Beatrix was fond of white rats, her dislike of bedbugs underlines their negative image.

Bedbugs' close connections with class attitudes is demonstrated in the story of Joshua Bug, a Yorkshire tavern proprietor who in 1862, presumably after a lifetime of ridicule, changed his last name to Norfolk Howard. Unfortunately for him, this rechristening did not save him from scorn. Newspapers had a field day with Bug's affectation in adopting the name of a premier noble family going back to the Middle Ages. The *Northampton Mercury* mocked, "To be ashamed of his own name is bad, but to affect a new high-sounding title, is atrocious vulgarity." From then until the early twentieth century, at least in some parts of Great Britain, bedbugs were referred to as "Norfolk Howards."[44]

In battle with the Norfolk Howards, the housewife wielded the weapons, especially if she wanted to secure the battlefield of the home. As we saw in chapter 1, cleanliness was becoming mandatory for modern middle-class respectability. But the labor of barring bedbugs from the home was enormous. Housewives had to air bedding regularly and dismantle beds as often as once a week. If bedbugs were found, their work increased exponentially, as this letter from Jane Welsh Carlyle to her husband reveals:

> I flung some twenty pails full of water over the kitchen floor in the first place to drown any that might attempt to save themselves—then we killed all that were discoverable and flung the pieces of the bed one after another into the great tub full of water—carried them up into the garden and let them steep there two days—and then I painted all the joints—had the curtains washed and laid by for the present, and hope and trust there is not one escaped alive to tell—Ach gott—What disgusting work to have to do—but the destroying of bugs is a thing that cannot be delegated.[45]

In 1916, the pioneering American entomologist C. L. Marlatt, whose major claims to fame were introducing the ladybug to the United States and charting the periodicity of cicadas, reflected on housewives' bedbug burden: "A Strenuous Struggle, a vigorous campaign, is before any housewife who is called upon to dispute the occupancy of her home with that persistent pest unfavorably known as the bed bug, who, gorged with the blood of his victim, lieth up in his lair from daylight to candlelight, only to swoop down upon his helpless sleeping prey during the midnight watches."[46]

The housewife was not without help, however. From the mid-nineteenth

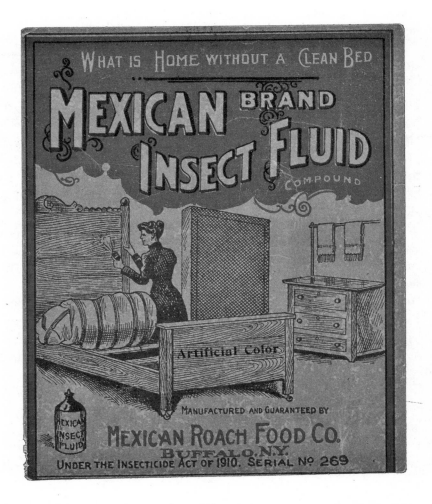

Insect fluid label, circa 1910–1915. Courtesy of the Science History Institute, Philadelphia, https://digital.sciencehistory.org/works/m039k510c

century until the present, companies selling bedbug products have invoked mothers and housewives protecting their children and homes. An advertisement by the Mexican Roach Food Company from the early twentieth century is emblematic of the connection drawn between good housewives and bad bedbugs.

This ad also reflects the continuing racism associated with vermin. Although the company was apparently founded to battle *la cucaracha*, by the early twentieth century, it was expanding its reach into the bedbug domain. Americans were thinking a lot about Mexicans in 1910 because of the Mexican Revolution and increasing Mexican migration into the United States. A company dedicated to destroying roaches and bedbugs would benefit from linking to attitudes about Mexicans.

The Mexican Roach Food Company was authorized by the US government, which was committed to licensing pest companies after the passage of the Insecticide Act in 1910. By 1912, the government was fining companies claiming to be able to eradicate vermin but actually selling specious insecticides.[47] But the government did not move to regulate sanitation in buildings, although the unhappy occupants of tenements pressed for action.

In the United Kingdom, the government did monitor bedbug incursions in buildings and took action against the infestations, with attitudes about class a key component of the effort. In the 1930s, at a meeting of the Medical Officers of Health in Great Britain, the president of the society, C. Killick Millard, MD, CM, announced that "the tenants of bug-infested houses are not usually the cleanest of people; but often they are to be pitied as much as blamed" and "that many of the class we are considering [people living in slums] have been accustomed to the presence of bugs, more or less, all their lives, and familiarity has therefore bred indifference." The answer to the problem of bedbugs, he suggested, was to educate the poor to "an anti-bug conscience."[48]

The effort led to the establishment of the Committee on Bed-Bug Infestation in 1935 by the British Ministry of Health. Its 1940 report included a section on Glasgow's work to eradicate bedbugs, or "slum-bugs," as one of its authors later called the insect.[49] The government had been relocating slum dwellers to new housing in that city, but its efforts would be of no avail without "Adequate supervision of such houses, together with instruction of housewives in simple methods of domestic hygiene." Even using hydrogen cyanide and sulfur dioxide to fumigate houses wouldn't work because "the use of chemicals in clearing a house of bed-bugs may result in tenants adopting an attitude of indifference to their responsibilities in maintaining such a standard of cleanliness in their houses as will ensure immunity from re-infestation. . . . There can be little doubt that the lack of domestic cleanliness is the principal factor in bed bug infestation."[50]

Government attention to bedbugs continued during World War II. A government report in 1942 noted that "In practically every community throughout Great Britain and elsewhere there are bug-infested dwelling houses which need not have been allowed to degenerate into this deplorable condition had an efficient system of supervision been exercised by those responsible for their maintenance." The report recommended, in language dripping with condescension, that sanitary inspectors emphasize the importance of cleanliness to housewives in a way that would not offend "the backward and lazy tenants. . . . An officious attitude is bound to fail, whereas a sympathetic though firm handling of even the most careless tenants will ultimately in most cases gain their willing co-operation." If the homes of such people could ultimately not be cleared of bedbugs, the report echoed the advice of the early exterminator William Cauty: "demolition of the building may be the only effective remedy."[51]

As we've seen, upper-class condemnations of the bedbug ridden as hygienically indifferent were common from the nineteenth century on. With no sympathy for the social conditions that might have spurred bedbug infestations, the government put the burden on housewives. The Medical Research Council's report tries to downplay the amount of work involved for the housewife to secure her home from bedbugs: "It is common knowledge that where there is cleanliness there are no bed-bugs, and it is also a fact that the degree of household cleaning required from the housewife to achieve complete immunity from the bed-bug for her household need not be of that striking degree which may even be uncomfortable."[52] Thus there would be no undue difficulty for the housewife in preventing bedbug bites. Unfortunately, this sanguine assessment was rendered obsolete as the Blitz forced Londoners into bomb shelters, where their bedbugs also sought sanctuary. The Ministry of Health decided it was imperative to publish the 1942 report because "the matter is of particular importance under war conditions."[53] It became a question of national security when bedbugs threatened not just the poor but everyone taking refuge in shelters. Fortunately, DDT, widely available at the end of the war, was a lot more effective than the products produced by the Mexican Roach Food Company. Here was a remedy to give all housewives some rest.

The Modern Bedbug Panic

The bedbug break ended in 1972 when DDT was banned. Several decades later, the creature reappeared, in a terrifying new onslaught. By this time, it was resistant to almost all forms of insecticide, creating, in the words of the entomologist Michael Potter—guru of bedbug exterminators—the "perfect storm" of pest control. An impassioned advocate of killing bedbugs, Potter

Penn Salt Chemicals advertisement extolling the virtues of DDT, *Time Magazine*, July 30, 1947. Courtesy of the Science History Institute, Philadelphia

chastised environmentalists for their indictment of pesticides: "I'd like to take some of these groups and lock them in an apartment building full of bugs, and see what they say then."[54] Bedbugs, it appears, have become another sticking point—or crawling point—in today's culture wars.

The scope of the current bedbug return is daunting. By 2011, a report by the National Pest Management Association reported that bedbugs had been found in 30 percent of public housing, and that 99 percent of pest control companies had been contacted to fight the menace.[55] Scientists face considerable problems when they try to devise new strategies for controlling and ending the onslaught. The bedbugs stocking research labs are bred from stock that is thirty years old and less pesticide resistant than the current breed—it takes a much larger dose of poison to kill bedbugs in the wild compared to their tamed brothers and sisters.[56]

And, of course, in this case the wild is your bedroom. "Often people don't want to use their house as a laboratory," notes a Finnish researcher. "They just want to get rid of them as quickly as possible."[57] Still, research continues. Scientists have sequenced the bedbug genome and determined how the creatures have outgrown vulnerability to pyrethroids, the most common insecticides. Some researchers have suggested that using bedbug pheromones may be the key—bedbugs reproduce by a method called "traumatic insemination," in which the male bedbug burrows a hole in the female, who has no genitalia of her own, and injects his sperm into it. (Yes, this is another reason that bedbugs are so horrifying; human beings who feel that their intimate spaces are being violated can relate.) The males are not discriminating in their sexual advances, which include attacking other males and immature nymphs, resulting in their deaths. These potential targets manufacture pheromones to let the males know they are not open for business, or open to being opened. If scientists can isolate these chemicals, they can use them to produce effective insecticides—or spermicides—to lure bedbugs onto traps with desiccant dusts, drying them to death.[58]

Naturally, as with many advances in science, there is a downside to dousing bedbug sexual relations. Bedbugs could be a tool to combat human criminals; forensic experts suggest that the blood that bugs have sucked up could be used in DNA analysis. And they provide employment opportunities for many people—and dogs.[59]

Less practically, but perhaps more profoundly, bedbugs may offer a way of understanding evolution. They are "living fossils" currently mutating into a completely new species as they become resistant to pesticides. Warren

Booth, a biologist at the University of Tulsa, writes in a study he coauthored in the journal *Molecular Ecology*, "For something that is so hated by so many people, it might just be a perfect model organism for evolutionary questions."[60]

But if bedbugs don't carry disease and their bites are relatively harmless, why do people hate them so much? Partly, it's because they attack us when we are asleep, at our most vulnerable. Bedbugs have resurged on the rising wave of international travel and globalization. They tag along on our luggage and backpacks, crossing borders in a way that would make anti-immigration folks crazy. Indeed, a guest at the Trump National Doral Hotel in Miami, staying in the Jack Nicklaus suite, sued the hotel chain for bedbug-inflicted injuries; the injured party declared, "Trust me when I tell you it is horrible to wake up in the morning and understand that bugs were crawling all over you all night. . . . I was deceived by Trump's image."[61] Not surprisingly, the case was settled before the 2016 election, for an undisclosed amount.

Bedbug lawsuits have created a lucrative legal specialty. A suit against a Red Roof Inn in Oxon Hill, Maryland, produced a judgment of $100,000. The attorney involved, Daniel Whitney, is so successful he is known as the "Bed Bug Lawyer." He developed the specialty after a scare that his daughter had brought the insects back from Europe. After an expensive treatment of his home, he had a revelation: "Wow. This has some real liability implications."[62]

The personal injury attorney Alan Schnurman sued New York's legendary Waldorf Astoria, describing his clients' experience as "the stay from Hell."[63] After one bedbug sighting, a hotel executive in San Francisco paid $2,500 to have his rooms steam cleaned. "It sounds like a lot of money, but the value of a good reputation is infinite," he told the *New York Times*. "Your biggest fear is that someone will get bitten and post something about it on an online travel site, and that'd be a killer."[64]

Dell Law Nationwide Headquarters, located in Miami, has a Bed Bug Division, which represents angry guests and tenants across the nation, with a twenty-four-hour hotline and a YouTube video advising what to do when bedbugs attack. The firm warns that bedbugs can leave scars, both in body and mind, and assures potential clients that there is no charge without a settlement.[65] Another bedbug firm, Bed Bug Law, uses rhetoric that sounds like a clip from the television show *Law and Order*: "Through our efforts we have earned our reputation as talented and courageous attorneys that possess the proven skill to pursue bed bug lawsuits and claims from start to finish. We work passionately and diligently to hold negligent business owners responsible for their actions and harm they have caused our clients."[66]

The profitable pursuit of bedbugs is not limited to lawyers. Since the mid-nineteenth century, when one inventor advertised his "Cochran's Improvement in Fastening Bedsteads" as a device to evict bedbugs, and Thomas Miller devised his bedbug exterminator, the tiny creatures have presented a huge business opportunity. There are products to prevent bugs from scaling the legs of a bed (ClimbUp Insect Interceptor) and others to encase a bed entirely (BugLock Bed Bug Mattress Encasement); there are special vacuums for bedbug removal (the Cimex Eradicator) and heating devices to kill bedbugs on luggage and backpacks (the Armato 9000), not to mention kits that help identify bedbug excrement (Bed Bug Blue Fecal Spot Detector). A company called PackTite, owned by a Colorado inventor, David James, offers products ranging in price from $19.95 for a bedbug monitor to $3,861.00 for a heater to kill house-sized infestations. His inventory moved him to poetry:

> Roses are red
> Violets are blue
> PackTite kills bed bugs
> After confirming fecal with Bed Bug Blue.[67]

Should all these products seem too mechanical, there are other bedbug strategies. An environmentally conscious company called Terramera sells a system called Rag-in-a-Bag, where rags treated with oil from the Indian neem tree are placed in a bag with infested possessions for a week. The company explains that a "strong natural green almond smell" will tell you it's working—ecological evidence evoking the strong smell approach of eighteenth-century exterminators.[68] The National Pesticide Information Center, by contrast, compares neem oil's smell to "garlic/sulfur," which might impress John Southall even more but be less seductive to modern noses.[69]

And if neem oil still seems too tame, bedbug–sniffing dogs and handlers are available, costing as much as $3,500. As with many bedbug products, there are no guarantees—sometimes the dog has not been trained properly, and sometimes it identifies false positives, particularly in the case of unscrupulous trainers.[70] But with apologies to the People for the Ethical Treatment of Animals (PETA), the idea of using a beloved animal—Beagles, German Shepherds, and Labrador Retrievers are best—to pursue a horrifying one is somehow validating.

The Beagle Roscoe who works for Bell Environmental Services and is described as Roscoe the Bed Bug Dog and his peers rely on smell to identify their quarry.[71] In earlier times, bedbugs' aroma was likened to coriander. Nowadays, it calls up different nose news. The EPA describes it as "mostly sweet and raw beef-like," while to the British Pest Control Association, bedbugs smell like "a mixture of raspberries, almonds and mouldy shoes."[72] These Brits are probably sniffing the pheromones emitted by female bedbugs to discourage the rapturous advances of the traumatically inseminating males. Swedish scientists are trying to inhibit bedbug procreation by covering bedbug sweat glands with nail polish.[73] The Harvard entomologist Richard Pollack says there has to be a whole lot of bedbugs for anyone who is not a professional to notice their smell, so perhaps it's better to leave it to the dogs.[74]

If you can't afford either mechanical or natural treatments, Bed Bug Pajamas Inc. will sell you a top-to-bottom protection unit. If the pajamas call to mind the hazmat outfits donned by people treating chemical spills and the coronavirus, it does not seem a coincidence. Like the earth-scorching DDT, the campaign against bedbugs can resemble a full-scale war of one species (*Homo sapiens*) against another (*Cimex lectularius*). Many of the bedbug products have names and taglines that echo the metaphors of battle: Armato—The Ultimate Bed Bug Killer, ClimbUp Insect Interceptor, EcoRaider, Bed Bug Patrol.

But in their sales strategies, advertisers often appeal to sentiments maternal rather than martial, to the protective instincts of mothers and their role in guarding the health of children and the sanctity of the home. Their advertisements tug at the heartstrings—one ad for EcoRaider Bed Bug Killer shows an adorable sleeping baby with the caption, "Sleeping Tight, Starting Tonight." No mother could resist it.[75]

The gendering of bedbug products—munitions to kill them and domesticity to protect against them—demonstrates both the psychological sophistication of advertising agencies and the money-making potential of the pest. AllerEase, a New York company directed at the bedbug market, is proud to be a women-owned company, showing bedbug defense as a female priority. Although not necessarily gendered, the company Ponctuel Escargot offers a "Chic Bedbug Jewelry Chain," perhaps seeking to cash in on the inherent absurdity of bedbug commercialism.[76]

Psychology and Society and the State

Driving the booming market for bedbug attack and defense is a simple truth: bedbugs can make people crazy. The paranoia they induce—and the commercial opportunities they provide—has driven the infested to suicide, murder, and generalized despair. As one bedbug expert puts it, "for those in fragile mental states, the bugs are devastating."[77]

Many Internet sites and blogs are devoted to helping those suffering physically and mentally from bedbugs. Perhaps the most informative, at least until it closed down in 2017, was Bedbugger.com. Its forum had hundreds of posts, starting in 2008, about the nightmarish experience of thinking you have bedbugs, sometimes even without any evidence of an infestation. One person posted, "I believe I have PTSD that is bed bug related, I can't sleep without drugging myself with sleep aids, I can't stop checking my bed . . . I just feel like I've been damaged beyond repair and I often feel alone in my struggle." Another contributor confessed, "I was convinced that I was doomed forever, and cursed to have to deal with them for the rest of my life. I started getting really anxious, depressed, would sometimes go days without sleeping. . . . It was a really dark time in my life."[78]

Such phobias can lead to terrible actions. The last contributor, a teenager, left home to sleep in a park or at his friends' houses. His parents refused to believe him, and he is now estranged from them. At least he did not try to commit suicide like one poor woman with a history of bipolar disease and alcoholism; she left a note saying, "I just saw a drop of blood on my dressing gown sleeve and I am sure vampires are back and I cannot stand to live in fear of me being eaten alive . . . At the time of writing, I have swallowed a bottle of wine and two hundred pills and I feel nothing. I feel completely empty, it is unendurable."[79]

Suicide because of bedbugs is bad enough, but the insect drove another mentally fragile individual to matricide. A Minnesota man killed his mother because he believed that "there was no way out" after treating her apartment with pesticide which he thought then poisoned the Twin City water system. As "nobugsonme," the webmaster of Bedbugger.com, commented, "Desperate people do desperate things."[80]

The common thread in all the bedbug postings is the victims' shame. They are "damaged beyond repair"; they feel "dirty and unclean."[81] Like victims of vampire attacks, the bitten become in a sense vampires themselves, facing social paranoia and ostracism. They fear that if they report a bedbug infestation they will be kicked out of their homes or refused social services—and this has happened to some sufferers. "If you live in a city," explained the *Washington Post*, "the word 'bed bug' is like an icy dagger to the heart."[82] Regular vampires are killed with wooden stakes through the heart; apparently, bedbug vampires and their victims require stronger measures.

In 2010, the comedian Stephen Colbert asked Jerome Goddard, who is currently an entomology professor at Mississippi State University, whether "the tiny vampires"—that is, bedbugs—are really a threat or whether the current hysteria is just a product of "media hype." It's a good question, and certainly the media has benefitted from the recent excitement. Displaying that, *BBC Magazine* breathlessly told its readers, "Vampire fiction may be all the rage. But the true bloodsuckers after twilight are not charismatic updates of Dracula but tiny insects living in our mattresses, headboards and pillows. Yes, bed bugs are back and pest controllers are warning of a global pandemic."[83]

TV programs from the *Today* show to *Dr. Oz* to *Orange Is the New Black* have highlighted the bedbug invasion. A Chicago television station reported that a passenger on the CTA Red Line had found bedbugs crawling on his seat, a claim later retracted when the pests in question were identified as body lice—but not before the story was picked up by the *Chicago Tribune*.[84] Similar tales have appeared from Canada to Kenya to China. The *New York Times* has been hot on the bedbug story from the beginning, publishing more than a hundred articles in all the sections of the paper (including Modern Love) since 2006. The webmaster of Bedbugger.com concluded that bedbug stories are simply more sensational than tales of lice and sell more papers or get more clicks.[85]

The bedbug outcry is out of all proportion to their actual threat. It is not just that they attack the sleeping and vulnerable, or that some people experience allergic reactions and welts. The reaction to bedbugs reflects our fear of social isolation. The stigma of bedbugs essentially makes a person a pariah who may drag friends, relatives, and strangers into the pit with him. "Still, the worst thing about bed bugs isn't the bugs themselves, or even the painful bites," confided Tess Russell in the Modern Love column of the *New York Times*. "It's the way others react when you give them the news, all variations on the same theme: they tell you how sorry they are, and then they back away."[86] As we've seen, Amy Schumer can attest to losing social appeal thanks to bedbugs.[87]

Social isolation is more terrifying than any bug, classifying the victim as an undesirable, an "other" blamed for his own misfortune. The EPA has tried to lessen the phobic reaction to bedbugs by urging people not to panic or throw out all their furniture, and not to keep an embarrassed silence about their problems: "Open communications about bed bugs and infestations," the government advises rosily, "will foster collaboration to solve the problem, rather than assigning blame or fostering stigma about infestation." The government particularly recommends dealing gently with bedbug bulletins from classrooms: "Bed bug sightings at schools can help identify potential infestations in the community; however, this information is very sensitive because it has the potential to stigmatize students and their families and should be handled discreetly."[88]

But the government is sending mixed messages on the pest and how to deal with it. The EPA recommends Integrated Pest Management techniques, which use knowledge about pests and their habitats to reduce the risk of infestations and employs pesticides only as a last resort. These methods are "smart, sensible, and sustainable" and "are particularly effective for owners/ agents of shelters, some group homes and other housing accommodations for transient populations where the risk of bed bug introductions and subsequent infestations is high."[89]

In its 2015 *Collaborative Strategy on Bed Bugs*, the Federal Bed Bug Workgroup, created after the Second Bed Bug Summit, urged "schools, housing providers, social service providers, pest management firms, local businesses, law enforcement and local health departments" to be sensitive to "cultural considerations . . . for example, values, ethnicity, national origin, language, gender, age, education, mobility, beliefs, standards, behavioral norms, communication styles, literacy, etc. that potentially affect management efforts and recommendations." The EPA cites the goal of discrediting "myths linking sanitation, poverty and immigration status to bed bugs."[90]

Does this list of cultural considerations imply that some people or communities actually do not mind bedbugs? When it comes to bedbugs, what goes around comes around—the tactics employed against the pest in the mid-twentieth century were almost the same as those used now, and the condescension toward the infested is also familiar. Dini M. Miller, a professor of entomology at Virginia Tech, developed a plan for shelters in her state that includes posters placed in common areas to increase "bed bug awareness," a call that replicates the "anti-bug conscience" of C. Killick Millard in 1934. Moreover, Miller advises those threatened to limit clutter, "the bed bug's best friend. . . . Limit the belongings that clients are allowed to keep in the shelter building, particularly non-essentials like plush toys, pillows, books, knick knacks, electronic items, etc. This may be difficult when children are in the shelter, but it is essential for making the living quarters less bed bug friendly."[91] The fact that taking these steps may make the shelter less human-friendly is not addressed.

Class consciousness pervades all the discussions of bedbugs and how to avoid them. Poor people, including abused women and children—and their knickknacks and stuffed animals—may indeed harbor bedbugs, partly because pest control is expensive and their belongings have often been donated. But some exterminators suggest that the greater responsibility lies elsewhere in the social scale. Because the stigma of having bedbugs is so profound, wealthier people at first deny they have a problem, allowing the infestations to turn massive. They also own more stuff, giving the bug more places to hide. When they finally admit the terrible truth, they are protected from public shaming by pest control companies using unmarked vans and pledging secrecy. According to Marshall Sella in *New York Magazine,* "High-priced specialists are enlisted to quietly rid Dior couture gowns, Porthault linens, and Aubusson silk rugs of their insect invaders. For those who appreciate irony, and perhaps a touch of Schadenfreude, there is this: Long-held ideas about bedbugs and poverty aside, wealthy people may in some ways be *more* prone to infestation. Bedbugs are equal-opportunity pests."[92]

In the past and the present, bedbugs drew government action when the bugs surfaced in the homes of the well-to-do—where, Millard observed in 1932, "it is considered bad form to even mention the word in polite society."[93] The bugs' appearance at that level prompted the Bed Bug Summits, the modern equivalent of the British Officers of Health. Local officials know that wealthier constituents expect this problem to be dealt with. In 2010, Mayor Michael Bloomberg, the richest man in New York City, complained to City Council member Gale Brewer that "all my friends have bed bugs; what am I going to do?" Another councilwoman, Christine Quinn, "shouted from the steps of City Hall, 'To the bed bugs in the city of New York . . . Drop dead. Your days are over, they're numbered, we're not going to take it anymore.'"[94]

Such bravado is perhaps inspiring, but bedbugs are impervious to words. All the powers of government, business, and the scientific community combined seem almost helpless against this creature that strikes in the night. Other insect vermin can be vanquished, but bedbugs symbolize our helplessness against hostile nature. Indeed, they seem to counter everything for which Western civilization prides itself. They are the natives let loose on the civilized. They are the weak revenging themselves on their conquerors. As early as 1930, at the height of the Empire, a British basket maker called bedbugs "little anthropophagi," or cannibals.[95] In the *Wall Street Journal*, Ralph Gardner Jr. labeled them "little terrorists" and "Soviet-style bogeymen at the height of the Cold War, the Osama bin Ladens of our pillows and bedrooms and pillows."[96] Jim Shea, a humorist at the *Hartford Courant*, confided, "Well, the CIA will deny this, but again my sources say that extremist Muslim bedbugs have been captured trying to enter the country wearing tiny suicide vests."[97] Fear of bedbugs and of terrorism occurring at the same time—in the first decade of the twenty-first century—is not lost on these commentators.

The huge threat and the little nuisance share some common attributes. Both attack when least expected, and they stir a fear far beyond their actual threat. Americans are more likely to be struck by lightning than blown up by a suicide bomber, and even less likely to be killed or seriously sickened by bedbugs. Yet their terror changes social behavior and travel plans. The fear makes us shun particular groups and ostracize certain people who threaten our feeling of invulnerability. We set up elaborate surveillance processes to detect the threatening other, such as the EPA's "intensive inspections" for bedbugs in certain places and the Department of Homeland Security's "extreme vetting" of immigrants or refugees from certain countries.

It's not surprising that popular culture has deemed bedbugs the ultimate enemy. In 2011, DC Comics created the villain "Bed Bug," who used the insects as weapons against his enemies, a threat far beyond the powers of any human—except, fortunately, Batman. (Ironically, bedbugs are found on bats as well as people, owing to their common habitation in caves thousands of years ago.)

For some Christians, Jews, and Muslims, bedbugs are proof that the end of times has come. Jessica Goldstein of the *Washington Post* described her battle with bedbugs in biblical terms. "Is someone smiting me?" she asks the pest control guy, referring to the plague of locusts God sent to afflict the Egyptians in the Passover story. After learning the true identity of her pests and waging war against them, she describes her experience as "Bugmaggedon."[98]

Atheists, however, may believe that the "vile, evil little creatures" disprove the idea of a benevolent God.[99] Whatever one's theological beliefs, it is clear that no one—except some tolerant entomologists and a few prosperous exterminators—likes the creatures. One Texas exterminator, after getting a $60,000 contract to eradicate bedbugs from an apartment complex, "had to pull my truck over and do a happy dance."[100] The Museum of Natural History entomologist Lou Sorkin is happy to offer his arm to bedbugs for feeding, explaining the bites don't hurt, and the Mississippi insect expert Jerome

Goddard dismisses fear of bedbugs: they are just "doing their thing." When asked why he allowed bedbugs to feed on his face, Goddard replied, "Oh, you know, bug people are crazy."[101]

Bedbug nuttiness even infests entertainers. The actress and model Isabella Rossellini perhaps captured this manic behavior in a video on the Sundance Channel. Dressed as a bedbug, she shouts, "Chase me, mate with me, seduce me," and writhes in pleasure as she is traumatically inseminated by the penis of the male bedbug, proclaiming "he is so strong and sharp" while he ejaculates into the wound. On the *Daily Show* in August 2010, Jon Stewart suggested it might be the worst Isabella Rossellini ad ever—might the fragrance be called "Infestation"? And after watching Lou Sorkin feed bedbugs on his arm, Stewart inquired, "So, is there a Mrs. Mad Scientist?"[102]

Nineteenth-century entomologists might have agreed that their twenty-first-century counterparts are a little nuts, although the amateur naturalists of the period, like Mr. Tiffin, possibly might have shared the enthusiasm for the creatures. Certainly, exterminators throughout time would rejoice at the prospect of making money from the bedbug-hunting business. Housewives of the past, whose unhappy job included trying to protect their homes and children from bedbugs, embraced any product, then or now, promising to control the beasts. From John Southall up to the owner of Roscoe, the Bug-Hunting Dog, bug hunters have used advertising to find clients and make money. Roscoe and his fellow canines find bedbugs with their noses, but their owners do not seem to find bedbugs nauseating; it seems even the most basic of senses is shaped by cultural attitudes. No one in the eighteenth and nineteenth centuries seemed particularly upset about bedbug bites—it was the smell that got to them. Perhaps we value our skin more than people in the past did. The claim that the poor were indifferent or apathetic about bedbug infestations was a condescension born of rising middle-class standards of cleanliness, which in the twentieth century solidified to cultural imperatives. Social authority linked bedbugs with the classes that endured them.

Recently, the *New York Times* columnist Bret Stephens took offense when David Karpf, an associate professor of media and public affairs at George Washington University, called him a bedbug on Twitter. The enraged journalist replied, "I am often amazed about things supposedly decent people are prepared to say about other people . . . on Twitter. I would welcome the opportunity for you to come to my home, meet my wife and kids, talk to us for a few minutes, and then call me a 'bedbug' to my face. That would take some genuine courage and intellectual integrity on your part."[103] Stephens was clearly aware of the long and unpleasant history of bedbugs and the implication of the connection; it affronted his humanity and his position in society, and even his role as a husband and father. It was an incendiary insult.

A bedbug link infuriated an even more powerful figure than Stephens. When the claim of bedbugs at the Trump National Doral Miami in Florida hit the newspapers again, after President Trump had suggested it as the meeting place for a G7 conference, the president vented on Twitter: "No bedbugs at Doral. The Radical Left Democrats . . . spread that false and nasty rumor."[104]

The bedbug mocks human superiority. Above all others, the creature seems to rebuke modern pretenses. People pride themselves on being able to sleep and travel where they want, and to mold nature and the environment to their liking. Not so, says this tiny and apparently quite cunning insect: You may make fun of me, or try to use science and government to control me, but I will outwit you. Chemicals fail to kill me, dogs cannot find me, and all your machines cannot vanquish me. Mothers and housewives can try to banish me, but I can migrate from place to place and reoccupy the homes of the afflicted when expelled by the neighbor next door.

In the war of all against all, the bedbug, hiding in our laptops and stuffed animals, seems impervious to our attempts to vanquish it. Social class doesn't matter; the pariah bedbug spreads social ostracism, and it makes no difference if you live on Park Avenue, at a Bronx shelter, or in public housing in Great Britain. Instead of being battered by the many ways people try to kill it, the bedbug evolves, still feeding on human blood and human fears.

Human vulnerability makes us paranoid. In the post-9/11 world, ever sensitive to symbols of a world out of control, the bedbug is the most horrifying of insects. It may take another kind of pandemic—the coronavirus—to impede bedbug hysteria. As the *New York Times* columnist Jane Brody writes, "Given the drastic pandemic-induced reductions in travel, the chances of bringing home these uninvited guests have been greatly curtailed."[105] But we can be sure that once COVID-19 is defeated by a vaccine, the bedbug will resume its merciless march across our pillowcases and skin.

CHAPTER THREE

Praying Lice

Crawling into Religion, Science, and Sexuality

Once there was a Roman dictator who suffered a gruesome end. One seventeenth-century minister described the fate of the Roman tyrant, Sulla, in a manner calculated to discourage anyone from following his example:

> He kept company with Actors, Actresses, and Minstrels, drinking with 'em
> Night and Day: He also indulg'd himself in Unnatural Lust with Men, particu-
> larly *Metrobius* the Woman-actor;—These lascivious Courses brought him into
> a Disease which made his Bowels to fester, so that at last the corrupted Flesh
> broke forth into Lice in such prodigious quantities, that all the Hands he could
> employ were not able to destroy 'em, so that his Clothes, Beds, Basons, and
> Meat was polluted with that Contagion . . . so that, in fine, he who had led
> an unclean Life dy'd of an abominable and unclean Disease.[1]

The disease was "phthiriasis," the so-called lousy disease, which purportedly caused the host body to be consumed from the inside out by its own vermin, turning the flesh into lice.[2] This fate, according to the seventeenth-century author who translated this description from Plutarch, was "the just Judgment of God upon him."

There is actually no such thing as lousy disease. The closest modern analogue is "delusional bug syndrome," with the sufferer insisting he is infested with lice, worms, or other parasites—a psychiatric rather than a physical ailment. But it is not surprising that the legend declared that Sulla and many other tyrants were consumed by lice. Infestation was twinned with moral turpitude and unnatural lusts, which called forth the judgment of God. God works by mysterious and not-so-mysterious ways, and believers often thought the particular punishment fit the crime.

As parasites infesting the body, particularly the hair and skin—the parts most linked with moral and physical decay—lice revealed both bodily and spiritual collapse. They provided earthly punishment for moral lapses, foreshadowing the retribution to come in hell. If nothing else could touch the depraved—especially the powerful depraved while alive—God could use this parasite to torment the wicked.

So, in the past, lice crawled not only through entomology but also through politics, morality, and religion. Even in the first scientific investigation of these insects in the seventeenth century, the naturalists' vision was colored by the tiny creatures' legendary links to human sordidness and vice. The English experimenter Robert Hooke, after looking at a louse under the newly discovered microscope, described it as a "saucy" creature that "skulks" in the private parts of human beings. Pornographers, also quite interested in empirical observation, capitalized on the throngs of lice inhabiting whores and their clients. Whether references to lice were pious or profane, they are a key to thinking in the premodern world.

The louse has been humankind's close companion—and intimate enemy—since we emerged as a species. Humans scratch at three varieties: head lice and body lice (*Pediculus humanus*), which live on the head and in clothes, respectively, and pubic lice (*Phthirus pubis*), which reside in the groin. Each needs the human body's warmth and blood to survive, and therefore the louse has kept up with humans' evolutionary changes, leading scientists to begin studying lice DNA for clues to primate evolution. Human head lice and chimp head lice are similar, pointing to a common ancestor before our two species diverged about 6 or 7 million years ago. Sometime around 3.3 million to 1.8 million years ago, humans started to lose their body hair, driving the louse to lodge on our heads. When our ancestors started to wear tight-fitting clothes, sometime between 500,000 and 72,000 years ago, some lice migrated into the brave new world of clothing, gradually growing bigger than their hair-confined cousins. Meanwhile, sometime during the last 7 million years, the pubic variety evolved on gorillas into an entirely different species. Scientists speculate that some hapless human got much too close to a primate relative (as the *New York Times* science writer Nicholas Wade advises, "Don't even think about it") who gave him or her an active souvenir of their union.[3]

In any form, lice are not like the benign bedbug, which carries no epidemic diseases. Rather, lice pass along typhus, trench fever, and relapsing fever, a role finally recognized in the early twentieth century. No one knew that lice distributed these diseases until the French bacteriologist Charles Nicolle identified the source of typhus as louse excrement in 1907, winning the Nobel Prize for his efforts. By 1935, in a not entirely unsympathetic account of the louse's role in the propagation of typhus ("In his unrestrained simplicity, he [the louse] is much like Rousseau's noble savage"), Hans Zinsser credits many of the major events of European history with the typhus bacteria sickening armies.[4] Amy Stewart picks up this theme in a recent book in which Napoleon's disastrous campaign in Russia is blamed on lice and typhus.[5] These medical facts would have astonished the physicians and scientists of the past who followed Aristotle in believing that lice were generated by sweat and had no medical association except perhaps as a cure for jaundice in children.

Today, our primary reactions to this particular parasite are disgust and fear of social ostracism. Confident in our power over disease, we still resemble our premodern ancestors in seeing lice as a moral rather than physical affliction. Too many parents have received the dreaded notice of lice infesting elementary school students' supposedly pristine hair. (Literally too many, as the presence of lice is greatly exaggerated in current society.) We don't think about God or kings or whores when we think about lice, but we do think about how lousy life can be. Lice are a small reminder that human dominance and presumption can sometimes hang on a hair. We want to think that only dirty people get lice—until they land on our own well-groomed child.

It is a commonplace of modern society that our ancestors lived lives that were "nasty, brutish and short." When the political philosopher Thomas Hobbes coined this phrase in 1651, he was describing the condition in the state of nature, before the creation of the state. For most of us in the present, the entire past is a kind of state of nature, not because of the absence of political authority but because the hygiene was so bad. A recent cartoon in the *New Yorker* captures our attitude about the past.

Before 1900, people may not have known much about lice, but they constantly felt their presence. They tried to rid themselves of lice, not necessarily because they wanted to be what we would think of as clean but because they just wanted the itch to stop. Cleanliness during most of this period did not include the body hidden under clothes or hats; in fact, undergarments were not worn by women until the eighteenth century. Hair and body were rarely washed because water was considered dangerous—as indeed it usually was. Sometimes, people tried to brush their lice off their bodies, and they sometimes used mercury and salt to get rid of vermin. The lice comb was a toiletry article for all classes, as shown in a 1598 painting by Caravaggio that depicts the well-dressed former prostitute Mary Magdalene with a lice comb

"Yes, it's a golden age—or would be, if we weren't all swarming with lice."

CartoonCollections.com

New Yorker cartoon depicting the modern-day view of hygiene in the Golden Age. Courtesy of David Borchart, CartoonCollections.com, CC135410

on her dressing table—in other words, she still suffered from vanity and pride, from which, presumably, only Christ could save her.[6] Lice were God's weapon against the unclean, but unclean often meant morally wanting rather than physically unkempt. And immorality could be found in all classes.

In premodern times, society was organized into orders, arranged hierarchically in descending levels from kings to nobles to merchants to peasants to beggars, a form of structuring often referred to as the "Great Chain of Being." The natural world also had its hierarchies; lions and elephants were at the top, and insects were at the bottom of the animal scale. There was an ordering even among insects; highest was the bee (often represented as a king bee), and at the very lowest level were verminous creatures such as lice, fleas, and worms (then always classified as insects). Following an analogical way of thinking, the best of each class was equivalent, as was the worst. So, to have lice meant not only that you might be dirty, but also that you were the lowest of the low, both a moral and physical pariah. Thus when a tyrant was consumed by lice, he was essentially dissolving into the least and most loathsome of creatures—he was being eaten by his social inferiors. Likewise, if a scientist spent his time looking at lice, he risked sharing the status of his

Michelangelo Merisi da Caravaggio (1571–1610), *The Conversion of the Magdalen*.
Detroit Institute of Arts, Gift of the Kresge Foundation and Mrs. Edsel B. Ford /
Bridgeman Images

subjects—many of the satirical attacks on the experimental or "new science"
of the seventeenth-century Scientific Revolution emphasized its association
with menial work, which only the lower classes performed, and with lowly
insects. Similarly, someone contracting lice from a sexual partner risked being
associated with peasants or beggars. Lice could turn the world upside down,
destroying borders between the classes that had seemed impenetrable.

In printed material in the mid-seventeenth century, particularly in religious works, references to lice surged dramatically. The increase was particularly striking between 1615 and 1653, a period of increasing political and
religious turmoil in England that included the reigns of the early Stuart kings
James I (1603–25) and Charles I (1625–49), the English Civil War (1642–49),
and the military rule of Oliver Cromwell (1649–59).

Preoccupation with lice displayed the parasite's role as a master metaphor
for the disintegration of the body politic. The king was the head of the state

and the church, and that state was coming apart. Charles I was careful to tuck his hair into a cap before it was separated from his body; accounts of his execution do not mention whether head lice rolled out as well. But John Pym, a leader of the parliamentary opposition to the king, was rumored by royalist supporters to have been consumed by lice. He wasn't, but it made perfect sense to his political enemies that bodily corruption should reflect Pym's corrupting of the state.

The inescapable bite of lice meant many things to the unfortunate people they inhabited: divine weapon, political scourge, scientific object, sexual companion. People who saw God's messages and looming omens in all aspects of their lives found inescapable meanings in the lice crawling over them. They recoiled when the new microscopes provided close-up views of the creatures, turning symbolic menaces into actual monsters. It would take another hundred years after the invention of the microscope, and a completely different taxonomy of the natural world, before the louse began to lose its multiple meanings and become domesticated as simply another kind of insect, albeit a disgusting one. It regained its previous reputation as a threat to humanity only when its medical role was recognized. At any level of scientific understanding, lice retained the capacity to get under people's skin—literally and figuratively.

Theological and Regal Lice

Lice were scriptural creatures. The most famous reference to lice, one with which any Briton would have been familiar, comes in Exodus in the King James Bible. Lice are the third of the ten plagues God sent against the Egyptians after the pharaoh refused to let the Israelites go: "And the Lord said unto Moses, Say unto Aaron, Stretch out thy rod, and smite the dust of the land, that it may become lice throughout all the land of Egypt. And they did so; for Aaron stretched out his hand with his rod, and smote the dust of the earth, and it became lice in man, and in beast; all the dust of the land became lice throughout all the land of Egypt" (Exodus 8:17). Perhaps this incident spurred ancient Egypt's invention of the lice comb. The Egyptians were renowned for personal hygiene, so this fine-toothed lice comb may show that the pharaohs had to deal with a lice problem even after the Israelites were allowed to leave.

Lice combs appear not only among the relics of ancient Egypt, but also in archaeological sites in Greece and in the farthest corners of the Roman Empire. Lice reportedly tormented tyrants in these areas, according to ancient

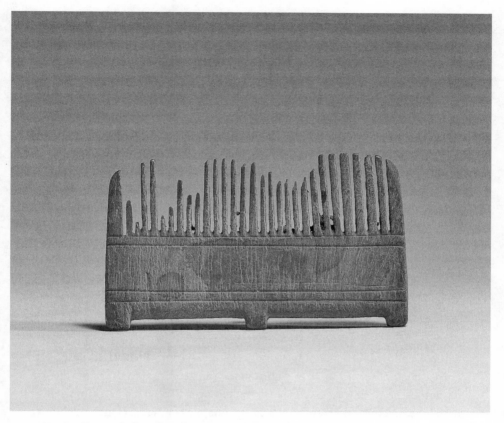

Wooden lice comb from Egypt, 1550–1307 BCE. Walters Art Museum, #61306

accounts read avidly in early modern England and British North America. In antiquity, lice afflicted not only the pharaoh but also Herod Agrippa and Antiochus Epiphanes, as well as the unfortunate Sulla.[7] Even those without royal credentials could suffer from lice—the philosopher Plato was supposedly consumed by them, perhaps as an ironic counterpoint to his philosophy, which privileged ideas over material being.

Well into modern times, lice were seen as a divine weapon eating into pride and sin. The religious worldview, today often considered antithetical to science and problematic in politics, was integral in the past's attitudes toward nature. Thus the seventeenth-century naturalist we met in chapter 1 opining about bedbugs, Thomas Muffet, called lice "the scourge of God." The Puritan clergyman Thomas Beard agreed, recounting the unhappy fate of the third-century Roman emperor Maximinus II (308–13 CE), whose "sicknesse was thus: In the privy members of his body, there grew a sudden putrifac-tion, and after in the bottome of the same a botchy corrupt bile, with a fistula, consuming and eating up his intrails, out of which came swarming an innu-merable multitude of lice, with such a pestiferous stinke, that no man could abide him." Beard adds that when Maximinus realized "that his disease was sent of God, hee began to repent the cruelty which he had shewed the Chris-tians, and forthwith commanded all persecution to cease."[8]

Of course, Christians as well as their persecutors had lice; even saints suf-fered from their depredations, requiring a modification in lice's theological mission. Although their presence was always physically unpleasant, spiritu-ally they could be interpreted as quite a good thing. From the Christian point of view, some writers argued, lice were a sign of holiness rather than a pun-ishment for sin, because any suffering evoked the suffering of Jesus. Some medieval saints, especially Franciscans, demonstrated their holiness, humil-ity, and imitation of Christ by embracing lice. Mortification of the flesh was achieved by a creature that lived on or under the tortured skin itself; this probably saved money on whips. The most famous medieval example of lice serving to sanctify through suffering was Thomas Becket, who as archbishop of Canterbury refused King Henry II's demands for control over ecclesiasti-cal law and property, leading to Becket's martyrdom in 1170. As Becket's corpse cooled, according to one chronicler, the lice living in the many layers of clothes he wore over a hair shirt "boiled over like water in a simmering cauldron, and the onlookers burst into alternate weeping and laughter."[9] Pre-sumably, the reaction of outsiders depended on their position in the battle between church and state, but it might have also testified to the spectators' joy at the apotheosis of flesh to sanctity.

Tyrants and saints weren't the only ones to suffer from this tiny tormenter. Lice plagued people from all walks of life—evil or good, sinner or saved. They seemed the primordial example of the theological challenge, "Why do bad things happen to good people?" The easiest explanation was that even the good were not without sin, so anyone infested with lice deserved it. Muffet explained that lice inhabit men because of original sin. Until Adam and Eve were corrupted by Satan, they were free of vermin, "but when he was se-duced by the wickednesse of that great and cunning deceiver, and proudly affected to know as much as God knew, God humbled him with divers dis-eases, and divers sorts of Worms, with Lice, Hand-worms, Belly-worms, oth-ers call Termites, small Nits, and Acares [mites]."[10] Once again, God used lice to punish sin, particularly sexual sin. Adam's legacy is lice as the scourge of all humans, not just the tyrants among us.

Catholics and Protestants, conflicted on so many things, agreed that human beings deserved their lice. The universality of vermin trumped any particular religious faith. Thus the Flemish mystic Antoinette Bourignon (1616–80) argued that we must accept with resignation the penalties that God has seen fit to send us for our sins. "All our Miseries and Corruptions come by Sin," she declared, "even the Vermine of our Body, the Lice, and all the other Beasts engendered by Corruption, are produc'd by Sin; and therefore we must suffer them willingly, since this is the Penance which God has in Justice permitted to befal us instead of the Punishments due to our Sins."[11] The minister Joseph Fletcher insisted in 1628 that God took just revenge on sinning man, "Sometimes by mighty troupes of Rats and Mice: And schoals of Wormes, and huge armies of Lice."[12]

The connection of lice with sin was a natural propaganda tool for the king's foes in the English Civil War, when the two sides were often distinguished by their hairstyles as well as their religious sympathies. The preacher Thomas Hall (1610–65), in his 1654 *The Loathsomeness of Long Haire*, urged "Go long-haired Gallants to the Barbers, go, Bid them your hairy Bushes mow, God in a Bush did once appear, But there is nothing of him here." He continued, "Tell me I pray, did you nere hear of Herods Executioner? . . . Or the third Plague of Pharaoh's Land?" Long hair and lice merged in the moral consciousness of the clergyman, who argued that long hair "doth but hinder men in their Callings many times: and without diligent care and much combing it becomes a fit harbor for Lice and vermin. Besides, some conceive that Long haire doth rather weaken than strengthen the body. . . . At best it is but a vaine and idle practice, of which men can give no good account."[13] Lice-ridden long hair will one day be judged by God, he warned, and the judgment will not be pretty. Some supporters of parliament were called Roundheads because of their short haircuts, while some who fought for the king were called Cavaliers because they embraced the long and flowing locks familiar to fans of romance novels. The hairsplitting debate ended with the restoration of Charles II in 1660, but as we will see, he may have had his own reasons to be worried about lice.

In the eighteenth century, when freethinking challenged the dominance of organized religion, the louse remained a source of theological conversation, although it was used to debunk religion rather than affirm it. A 1746 exchange in *Gentleman's Magazine*, the first general-interest magazine directed at the educated public, contains several letters on the burning issue of whether Adam and Eve had lice. One of the correspondents remarks that "We can hardly suppose that it was quartered upon Adam or his Lady, the neatest, nicest pair, (if we believe John Milton) that ever joined hands. And yet, as it disdains to graze the fields, or lick the dust for sustenance, where else could it have its subsistence?"[14]

It was a good question, especially if one believed that God created everything—including the "creeping" things—during the first six days. Edward Cave, editor of the magazine, suggested that lice did not come into being until God punished the pharaoh, but another correspondent replied that to deny lice's presence in the Garden of Eden essentially destroyed the legitimacy of Scripture. Accordingly, this supposed biblical literalist concluded that on the sixth day, God commanded, "the male and the female of this animal to bend their course towards Adam's head, where he order'd them to fix their habitation, seek their food, and increase and multiply." From thence they migrated to the head of "Mrs. Eve" and her progeny forever.[15] Perhaps this meal was bliss for the vermin, but it is difficult to understand how Eve might have enjoyed being dinner for a bug, although, as we will see later, some people found great pleasure in scratching their bites.

This clearly satirical debate ended with Edward Cave declaring that "the light of nature" will prove which biblical account contains the correct lice origin story: Moses or Mother. At any rate, according to him, the louse does serve a useful purpose in the scheme of things, although perhaps not from its own point of view. "Unhappy little creature," Cave bemoaned, "first occasionally created for the punishment of haughtiness and pride, and then forever sentenced to be the scourge of sloth and nastiness."[16]

Sloth and nastiness popped up in the most unusual places in the eighteenth century, including the royal court. Just as lice could undermine religious authority, they could also be used to undercut rulers who pretended to be above their people or thought of themselves as more than human, particularly in an age that saw the birth of modern democracy. Bedbugs plagued Princess Charlotte, the daughter of George IV, but a louse was the bane of her grandfather, George III, about whom a detractor inquired, "Maggots are found in many a princely head; And if a maggot, why then not a louse?" The poet Alexander Wolcott (under the nom de plume Peter Pindar) directed this query at the bedeviled George III, who among all his other troubles one day discovered a louse on his dinner plate. This event moved Wolcott to write a

146-page mock heroic poem on the subject, titled *The Lousiad* (1786–92), which begins,

> The Louse, I sing, that, from some head unknown,
>
> Yet born and educated near a throne,
>
> Dropp'd down, — (so will'd the dread decree of Fate,)
>
> With legs wide sprawling on the M—ch's plate . . .
>
> Yet, could a Louse a British King surprize,
>
> And, like a pair of saucers, stretch his eyes?
>
> The little tenant of a mortal Head,
>
> Shake the great Ruler of three realms with Dread?[17]

The appalled king, as offended by this act of *lese majesty* as by the machinations of his political foes, orders an inquisition of his household to uncover the head that produced this creeping threat. Ultimately, his cooks are ordered to shave their heads, to which they respond as rebelliously as had his former American subjects:

> What creature 'twas you fond [*sic*] upon your plate
>
> "We know not—if a louse, it was not ours:—
>
> To shave each Cook's poor unoffending pate,
>
> Betrays too much of arbitrary pow'rs . . .
>
> But grant upon your plate this louse so dread,
>
> How can you say, Sir, it belongs to us?[18]

Wolcott's satire, skewering every member of the royal family, attacks George III as an imperious but absurd would-be tyrant who manages to befoul everything he touches or that touches him, from the smallest (the louse) to the largest (America). Like rulers before him, George discovers that lice taste no difference between royal and ordinary blood, and once again they become the means for challenging royal presumption. By the eighteenth century, however, the louse is an independent agent rather than God's weapon against the evildoer. Nothing says more about the decline of royal and religious power in the modern age.

Scientific Lice

By the time Alexander Wolcott ridiculed George III and his louse, natural history—which would evolve into the modern disciplines of botany, zoology, and geology—was competing with physics, astronomy, and mathematics for status and royal patronage. All these disciplines had found a home in the

Royal Society, an institution devoted to scientific enterprises founded in 1660 that included among its luminaries Robert Boyle and Isaac Newton. Despite such distinguished scientists, the society attracted criticism from its foundation for a foolish devotion to useless facts and endeavors, a critique that intensified in the later eighteenth century during a period of comparative lack of vitality. Wolcott, perhaps because he was denied admittance to the society, was part of this tradition. Wolcott mentions his disdain for the Royal Society and its president, Sir Joseph Banks, whom he saw as a sycophant serving the interests of the commercial and colonial classes. In *The Lousiad*, he writes, "Then bid the vermin on the journals [of the Royal Society] crawl,/Hop, jump, and flutter, to amuse us all."[19] His satire shows that although amateurs and professionals had been investigating and collecting the objects of natural history for more than a hundred years by this time, there was still something suspect about this enterprise. Bugs came with a lot of baggage.

In fact, when it came to speculating about vermin, investigators of nature had to contend with not only lice's theological and tyrannical associations but also ancient and medieval philosophic and medical teachings about insects. In addition, lice—particularly in their pubic form—had sexual implications that sometimes colored the work of both the scientists and their critics. Experimental science and pornography shared an interest in the close observation of nature, sometimes wittingly and sometimes unwittingly. To some of its critics, the scientists of the Royal Society were voyeurs engaged in an attempt to strip and subjugate the natural world.

In science, the first—and for many centuries the last—word on the subject of lice came from the ancient Greek philosopher Aristotle, whose works were still taught in the universities of England until the nineteenth century. Aristotle believed that lice were generated by decaying flesh or sweat, the same origin attributed to bedbugs in the seventeenth and eighteenth centuries; they were a prime example of spontaneous generation, in which an animal is generated not by the sexual concourse of two similar parents but from another kind of organism altogether, or even from inanimate matter. Aristotle argued for the existence of an especially egregious form of "wild lice" that affected women and children particularly because they had a superfluity of moisture, which contributed to the generation of the insect.

Right up until modern times, the belief in spontaneous generation of vermin was pervasive, shaping medical care. Like Aristotle, Thomas Tryon, the late seventeenth-century advocate of clean air and clean bedding, thought that lice (and bedbugs) arose from sweat and filth. The seventeenth-century physician Nicholas Culpeper (1616–54), in a treatise directed at midwives and women, taught that since women and children "are hot and moist," they "have many excrements that are fit to breed lice." Expanding on his subject, Culpeper suggested that one way to prevent this is to "let children eat no food of evil juyce, especially Figs."[20]

The Dutch microscopist Antonie van Leeuwenhoek (1632–1723), who regularly conveyed his findings to the Royal Society, rejected such an explanation: "this notion took its rise from no other cause than the multitude of small grains and seeds with which all Figs abound . . . [and] it is very probable, that people comparing these seeds or grains in Figs, by reason of their smallness, with lice, first broached the idea that lice can be bred from Figs."[21] Such "idle tales," he commented, "certainly do no other than excite derision." The same kind of thinking caused lice to be considered a remedy for jaundice and eye complaints, perhaps because the yellowish insect was similar to the yellowed eyes of the patient. The therapeutic use of a similar-looking plant or animal, called the doctrine of signatures, was common in medieval and early modern medicine. It made eminent sense to many religious and medical thinkers who thought that God would have impressed the image of the antidote to any ill on the plant or animal that resembled the complaint.

But neither the doctrine of signatures nor spontaneous generation, nor any other folkloric tradition, made sense to Leeuwenhoek. The Dutch lens maker was the first microbiologist; he refined the recently developed microscope to the point that he could observe not only lice but also bacteria and spermatozoa among other objects not visible to the naked eye. Through these observations, he was able to establish, at least to his own satisfaction, that there was no such thing as spontaneous generation in nature and that everything in the created world demonstrated the omnipotence of God. Even insects, through their "incomprehensible perfection," demonstrated his providential design. One of these proofs of divine planning, he argued, is the louse whose "feet and claws . . . display the perfect contrivance manifest in the formation of so small a creature."[22]

Leeuwenhoek's admiration for the structure of the louse, however, did not prevent his disgust for the creature when it inhabited his body. And so he set out to study more closely "This animal [the louse], which is so troublesome to many, especially the poor, who have not the means of frequently changing their linen and other apparel," and which "some writers supposed to be produced from dirt, sweat, or excrements" but cannot be produced "otherwise than by the ordinary course of generation."[23] Leeuwenhoek made it his task to track procreation among the lice.

Leeuwenhoek was first to observe a louse's organs under a microscope, determining its anatomy and viewing the many eggs the female louse carried. Overcome with curiosity, the Dutch scientist also decided to test the belief that lice are "so prolific an animal, that it is a common saying, that it will be a grandfather in the space of twenty-four hours." At first, he was going to put several lice into the stocking of a poor child, but then he decided to use himself for the experiment—there is no data like the data produced by your own body, an attitude that continued until modern times. He placed two female lice in his stocking, securing them from escaping, and waited first for six days before examining the lice and the number of offspring they had produced: there were ninety. For the second experiment, he wore the stocking for ten days, producing two hundred and fifty eggs with many remaining to be born. But it all was too much for our intrepid experimenter: "But I was so disgusted at the sight of so many lice, that I threw the stocking containing them into the street; after which I rubbed my leg and foot very hard, in order to kill any Louse that might be on it, and repeating the rubbing four hours afterwards, I put on a clean white understocking."[24]

Ultimately, Leeuwenhoek determined "that two females may, in eight weeks' time, be grandmothers, and see ten thousand lice of their own offspring, which, unless reduced to actual demonstration, would seem incredible." He thus concluded, "we may easily conceive, that a poor person who has a hundred female Lice about his body, and has not any change of clothes or linen, and who, moreover, through sloth is careless of destroying those he has about him, may in a few months (if I may use the expression) be devoured by these vermin."[25]

There has hardly ever been a more poignant description of the plight of the poor, or a more empathetic understanding of their suffering. But even in the midst of his sympathy, there is a flicker of moral condemnation for those who do not keep themselves free of lice because of casualness with their underwear. The Dutch were famous in the sixteenth and seventeenth centuries for the cleanliness of their towns and streets. Reformers wanted to extend this fastidiousness to personal cleanliness, and they advocated the wearing of clean linen. With Leeuwenhoek, science, morality, and hygiene merged, anticipating what would happen in England, America, and the rest of Europe only much later.

Leeuwenhoek's lice findings were published in the *Philosophical Transactions* of the Royal Society, which elected him to membership in 1680, although many fellow academicians still clung to the doctrine of spontaneous generation in the insect world. The Dutch microscopist's entomological interests mirrored earlier work by the society's first professional experimentalist, Robert Hooke. Up until the mid-seventeenth century, there had been little interest in natural history in England, although Francis Bacon, the prophet of the "new science," had advocated its study. But by the time of the 1660 foundation of the Royal Society for the Advancement of Knowledge, experimenters paid increased attention to natural objects, perhaps in part because the invention of the microscope allowed specimens to be viewed more intimately and displayed more publicly. In the fifteenth and sixteenth centuries, Europe had seen the rise of a culture of collecting unusual and even monstrous natural specimens, from both the newly discovered parts of the world and home. Even kings displayed these objects in so-called cabinets of curiosities. When microscopes were aimed at prosaic objects like insects, they revealed spectacular and even horrifying features.[26]

And so, when the famous architect and scientist Christopher Wren, a founding member of the Royal Society, presented drawings of a magnified louse and flea to Charles II, the king was delighted by their strange and monstrous appearances and added the drawings to his cabinet. Perhaps he saw the louse subjected to the gaze of the observer as a symbol of royal control over nature and religion, transforming the creature that had been used to torment tyrants into a domesticated object at the mercy of the divine ruler. He commanded the Royal Society to continue the investigation of insects with the microscope, a task that eventually fell into the hands of Robert Hooke. A scientist of fairly humble origins, Hooke was an assistant to the noble experimenter Robert Boyle, and by 1665 he was the curator of experiments to the Royal Society, which hoped for continued support from the king. Royal patronage of this new institution may have hinged on the wing of a fly, or the bite of a bug.

Some students of nature contested this investigation of the miniature world. The new scientists had to justify their enterprise in terms of the traditional connotations of vermin. Why were they bothering themselves with these paltry and disgusting creatures? The physician Théodore de Mayerne, in his dedication to the second volume of Topsell's *Historie of Four-Footed Beasts and Serpents and Insects*, had faced this problem even before the microscope was used in Britain. He anticipated what would become the standard justification for the study of insects, the argument from design: "He [Muffet] thought it no indignity to Dedicate to the greatest Princess [his treatise on] the miracles of Nature, which are most conspicuous in the smallest things; which testifie [to] the infinite power of the supreme Creator of all things, and raise the minds of Princes who are children of the most Highest, to the cause of all causes."[27]

It is significant that so many natural philosophers, as investigators of nature were called at this time, included God as an element in their researches, revealing both the tenacity of traditional belief and the transitional nature of their enterprise. Like Leeuwenhoek, most considered it as only rational to see God's providential plan in the order of the universe. Many members of the Royal Society embraced the argument by design to prove God's creative power. It is anachronistic to see science and religion as antithetical before the eighteenth century—and, naturally, lice were embraced as (or at least considered to be) part of God's plan.

Vividly depicting this attitude toward God and nature was the naturalist John Ray's 1691 work *The Wisdom of God Manifested in the Works of Creation*. Although Ray and other theologians assumed God to be omnipotent, transcendent and beyond human understanding, they argued that "all the visible works of God" reveal "his Wisdom in the Composition, Order, Harmony, and Uses of every one of them." All creation, including the very large and the very small, "serve not only to Demonstrate the being of a Deity, but also to illustrate some of his principal Attributes, as namely his Infinite Power and Wisdom." For example, Ray argued, God made the "noisome and troublesome creature a Louse and put them in foul and nasty clothes . . . to deter Men and Women from Sluttiness and Sordidness, and provoke them to Cleanliness and Neatness."[28] His fellow member of the Royal Society, the physician John Wilkins, argued for a more positive role for vermin in God's creation, that even lice testified to divine providence and design: "Such accurate order and symmetry in the frame of the most minute creatures, a Lowse or a Mite, as no man was able to conceive without seeing them."[29]

The theological implications of studying the minute things of nature were graphically depicted in the signature text of the Royal Society, *Micrographia*, written and illustrated by Robert Hooke and published in 1665. Hooke argued, "And indeed, so various, and seemingly irregular are the generation or productions of Insects, that he that shall carefully and diligently observe the several methods of Nature therein, will have infinitely cause further to admire the Wisdom and the Providence of the Creator."[30] It is the "All-wise God of Nature" who disposes the little automatons of nature, the way a watch-

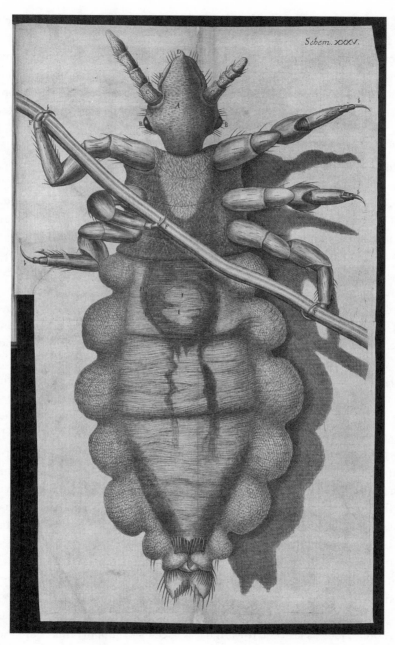

Robert Hooke, *Microscopic Image of a Louse*, engraving from *Micrographia*, 1665. Courtesy of the Wellcome Collection, https://wellcomecollection.org/works/hfmyu332

maker composes the watch from different materials. (This analysis reflects the doctrine of the French philosopher René Descartes, who thought that animals were automatons.) The insect automata of nature prove the ultimate design of God. Their static depiction on the pages of *Micrographia* demonstrates what we might call a theological moment in time, when creation is frozen for the edification of the viewer. The immobile flies, gnats, spiders, and even fleas of *Micrographia* are described as awesome and beautiful. In fact, the only insect Hooke seems to dislike is the louse, whose towering and intimidating image is the last of the large foldout engravings in *Micrographia*, reflecting its lowly status on the Great Chain of Being.

Lice were a problem for Hooke, and not just because he wore his natural hair long and lank. Unlike other insects, which were social (ants) or helpful to human beings (bees), or at least departed after biting (fleas), the louse gnawingly settled in its human host. Thus the louse exhibited in Hooke's *Micrographia* is attached to a human hair, the only creature in the book depicted in relation to man rather than by itself. Just like Leeuwenhoek, Hooke used his own body as a resource in his experiment, allowing a hungry louse to feast on his flesh and drink his blood. His description of the experiment is as much moral as scientific: "the Creature was so greedy, that though it could not contain more, yet it continued sucking as fast as ever."[31]

Hooke does not explain why God made lice so voracious, or what role they play in human economy, but he knew that they demonstrated human vulnerability. Hooke describes the louse as "a Creature so officious, that 'twill be known to every one at one time or other, so busie, and so impudent, that it will be intruding it self in every ones company, and so proud and aspiring withall, that it fears not to trample on the best, and affects nothing so much as a Crown."[32] Hooke often uses puns in *Micrographia*, and here he is playing on the double meaning of "crown," as both the top of the head and the mark of a king. Officious and impudent lice, insolently and disrespectfully, attack both the head of the body and the head of the state.

In airing the dirty linen of microscopic investigations, Hooke's description of lice was less reverent and more prurient than Leeuwenhoek's. Some modern feminist critics accuse the seventeenth-century experimenters of an attempt to exploit and even rape the natural world and force it into submission.[33] Hooke's attitude toward nature seems to support this contention. Knowledge of hidden things, he explained, had been formerly obstructed by nature, "who not only in her ordinary course, but when she seems to be put to her shifts, to make many doublings and turnings, and to use some kind of art in indeavouring to avoid our discovery."[34] In this description, "shifts" is both a verb and a noun; in the seventeenth century, it meant not only moving around but also the slip worn by women under their dresses. Hooke continues, however; once the microscope exposes nature's secrets, presumably both hidden and sexual, the result will be intense pleasure for the viewer, both mental and physical. He writes, "I do not only propose this kind of Experimental Philosophy as matter of high rapture and delight of the mind, but even as a material and sensible Pleasure."[35] Microscopy, it turns out, is not only a spur to religious devotion but also a sensual practice—and body could trump spirit in the pursuit of nature.

Thomas Sprat (1636–1713), writing a defense of the Royal Society in 1667, tried to cast the sensual pleasure found in experimentation as an innocent pastime. He asks, "What raptures can the most voluptuous men fancy to which these are not equal? Can they relish nothing but the pleasures of their senses? They may here injoy them without guilt or remors." Echoing Hooke, Sprat argues that the only things men need to enjoy the study of nature is "the use of their hands, and eies, and commonsense" to understand the matter in motion that constitutes the world. And thus, he concludes, even the great ancient hedonist Epicurus would have no reason to reject experimentation that opposes "pleasure against pleasure."[36]

These early scientists believed in using their hands to explore nature and perhaps themselves. Robert Hooke's graphic description of lice in *Micrographia* can be read as part of a tradition of sexualizing insects and their prey, particularly when description becomes practice in his work. "I found," he narrates, "upon letting one [a louse] creep on my hand, that it immediately fell to sucking . . . nor did it cause the least discernible pain." Did it cause pleasure? Hooke does not explicitly mention the pleasure that comes from scratching a louse bite, but he does allude to the sexual connotations of lice. Putting himself into the mind of his subject, he reflects that a louse "is troubled at nothing so much as a man that scratches his head, as knowing that man is plotting and contriving some mischief against it, and that makes it oftentime skulk into some meaner and lower place, and run behind a mans back."[37] The meaner and lower place is undoubtedly the private parts, where its cousin, the crab louse, is already in residence.

So, when Robert Hooke's rebel louse "feeds and lives very high . . . [and] affects nothing so much as a Crown," his words can be read as a gloss on the monarch's private parts as well as his kingly head. In implying that the king was infested with lice that "skulk into some meaner and lower place," Hooke perhaps inadvertently joined a chorus of detractors and satirists who linked Charles's sexual excesses with corruption. His audience certainly could have read his description of the louse as a kind of attack on royal authority. Hooke's contemporary Andrew Marvell advised a painter in a 1667 poem,

> Or if to score out our compendious Fame,
> With *Hooke* then through your Microscope take aim,
> Where like the new Comptroller all men laugh,
> To see a tall Louse brandish a white Staff.[38]

The louse in this poem is Lord Brounker, president of the Royal Society and newly appointed comptroller of the navy. The hair gripped by the louse in Hooke's illustration has metamorphosed into Brounker's staff of office. Its phallic nature also replicates the scepter carried by the king as a symbol of his power—sexual or otherwise. The king and his minister have become a laughingstock. A potential tyrant has been cut down to size by a louse. Hooke could not have been overjoyed to see his microscopic observations used so quickly to satirize the king whose support he sought—although he was not above making a dirty joke or two in *Micrographia*. For example, the itch of an insect bite is like "the stinging also of shred Hors-hair, which in merriment is often strew'd between the sheets of a Bed."[39]

At least one of his readers found Hooke's description of vermin and other objects less than amusing. Margaret Cavendish, Duchess of Newcastle, was the first Englishwoman to publish on scientific subjects, so her reflections on *Micrographia* in her scientific treatise *Observations upon Experimental Philosophy* (1666) and in her satiric romance *The New Blazing World* (1666) are especially revelatory. Cavendish thought that the observations of Hooke and the other members of the Royal Society distorted what they saw, creating what she called "hermaphroditical" objects that were monstrous combinations of nature and art. Hermaphrodites were considered unnatural in the seventeenth century and were paraded at fairs and circuses as "jokes of nature." They personified disorder in nature. Thus the observations of the experimenters are sexualized, and the explorers of nature are, in a sense, voyeurs of natural deviance—which, paradoxically, they have created. Cavendish makes this connection explicit in her criticism of Hooke's louse. She argues that a louse "by the help of a Magnifying-glass, appears like a Lobster, where the Microscope enlarging and magnifying each part of it, makes them bigger and rounder then naturally they are." The resulting images, like hermaphrodites, are unnatural: "The truth is, the more the figure by Art is magnified, the more it appears mis-shapen from the natural, in so much as each joint will appear as a diseased, swell'd and tumid body, ready and ripe for incision."[40]

The swollen louse is ripe for incision or penetration by the experimenter. Its plight is analogous to that of a young lady who is observed too closely through a microscope: Cavendish cries, "Nay I dare on the contrary say, had a young beautiful Lady such a face as the Microscope expresses, she would not onely have no lovers, but be rather a Monster of Art, then a picture of Nature." In fact, the young lady would "have an aversion, at least a dislike to her own exterior figure and shape; and perchance if a Lowse or Flea, or such like insect, should look through a Microscope, it would be as much affrighted with its own exterior figure, as a young beautiful Lady when she appears ill-favoured by Art."[41] (By "art," Cavendish here means microscopic observation.)

Young ladies have reasons to fear lice, and not only because they could share the point of view of insects at the sight of their own features reflected and exaggerated by the microscope. In Cavendish's science fiction romance *The New Blazing World*, her protagonist, the Empress of the Blazing World, reacts strongly to the sight of the engraved images of these insects. "Lastly," Cavendish writes, "they shewed the Empress a Flea, and a Lowse; which Creatures through the Microscope appear'd so terrible to her sight, that they had almost put her into a swoon."[42] The duchess's character may have almost fainted because she feared the intrusion of both the experimenters and the vermin they studied. In a sense, she is being ravaged by the scientists who are challenging her authority as well as her virtue. Not surprisingly, to prevent rebellion in her realm, she ultimately decides to dissolve all scientific societies.

Lice and the Body

Graphic imagery characterized both scientific and pornographic texts in the late seventeenth century, and both enjoyed a wide audience. Both shared an interest in revealing what had been previously hidden—in explicit detail. Not surprisingly, lice pop up in both narratives. The first European pornographer, the Italian Pietro Aretino (1492–1556), incorporated crab lice into his stories about prostitutes. *The Wandring Whore*, translated into English in 1660, includes a story about "a handsom neat Clean-skin'd Girl that was terribly pepper'd with Herds of Crabblice (as visible in her tayl as Cloves in a gammon of Bacon)." One does not need a magnifying glass to see the vermin, but their destruction involves a kind of experiment. A friend devises a way to rid her of her torment: "He took her and ty'd her up naked at the beds feet like a monkey, and lighting a pipe of Tobacco gave those many-footed vermin (arising from inbred Lechery) such a rout at her Cinque ports, by thrusting in the small end of his pipe into one hole, then into the other, blowing the smoke at the other end of his Pipe, that they never durst venture to inhabit those Continents since."[43]

Such treatment was presumably more literary than real. But homemaking manuals and medical treatises were full of cures for lice, which often included the same ingredients as treatments to destroy bedbugs. An early recipe suggested the following:

> Take the Dregges of Oyle, or freshe Swines Greace a sufficient quantitiye
> wherein you shall chase an unve [ounce] of Quick Silver till it be all suricke
> into the greace, then take powder of Staves-acre fearced and mingle all to-
> geather, make a Girdle of a woolen list for the middle of the pacient, . . . then
> let him weare it continually next to his skinne, for it is a singular remedy to
> chase them away.[44]

As late as the mid-eighteenth century, quicksilver or mercury was recommended as a remedy for crab lice, but the Italian doctor Vincenzo Gatti cautioned that a patient "found the Penus plainly cold and torpid, and quite unfit for Venery" after the treatment.[45]

An earlier writer, the Dutch humanist Daniel Heinsius (1580–1655), seized on the subversive and humorous possibilities of vermin, particularly when linked with sexuality. In an ode in praise of the louse, the anthropomorphized insect who narrates the piece suggests that lice do humans a favor because they provide the opportunity to scratch: "if there be any paine," he argues, from the itch provoked by lice, "it is the progenitor of pleasure, which dainty kinde of tickling (my Lords) I think you are so taken with, as that I imagine, it is your chiefest and most lushoius relishment or your poore and miserable condition. Often have I seene with what expressive delight, you use to rubbe and scratch, sometimes your head, sometimes your sides, sometimes another part, to which your guest gives the gentle itching twist."[46] The afflicted person experiences a "mighty fricative pleasure" after his itch is scratched. The sexual connotation is clear; the masturbatory function of lice emphasizes the illicit pleasure often connected with this insect.

In his ode to the louse, Heinsius was demonstrating the pungency of his wit while also stating a fact few liked to acknowledge—animals, including people, like to scratch. The *New York Times* recently reported that scientific itch researchers (yes, there is such a thing) have found that "itching and scratching engage brain areas involved not only in sensation, but also mental processes that help explain why we love to scratch: motivation and reward, pleasure, craving and even addiction. What an itch turns on, a scratch turns off—and scratching oneself does it better than being scratched by someone else."[47]

The humorous—and sexual—potential of scratching vermin, in this case mutual rather than solo, is caught in a long poem by William Cavendish, then Marquis of Newcastle (1592–1676) and husband of Margaret Cavendish, published in 1654. He describes the romance of an ancient beggar and crone, who "In a hot Summer's day, they out did creep, Enliven'd just like Flyes, for else they sleep; Creeping, at last each one to the other get, Lousing each other, kindly thus they met." Lice serve as foreplay in this account—picking lice off each other gets their "dead ashes" moving and they marry, although, unfortunately, age inhibits the consummation of their passion: "She, gentle Dame, with roving hand, indeed, Instead of Crutches, found a broken Reed."[48]

Yet another Restoration piece of pornography, written by Charles Sackville, 6th Earl of Dorset, describes the battle between two crab lice who live on the pubic hairs of a whore, to whom men scorned by lovers come to exorcise their grief. But as in all things, the poet comments, "And yet what corner of the World is found,/Where Pain and Pleasure does not still surround?" With the pleasure of finding relief in her "tuft," the clients also "fear their Ruin [in the] midst of their Delight."[49]

What weapons the raging crab lice use in their fight are unknown, unless "Some Greshamites perhaps with help of Glass,/And pouring long upon't, may chance to guess." The Greshamites are the members of the Royal Society who met at Gresham College in 1668 when this poem was written, and their glass is the microscope, now specifically linked to prurient activity. The lice only cease their fight when they fall into the whore's "briny lake," where perhaps not surprisingly they find "the trouble of a tott'ring Crown,/Of a mighty Monarch is laid down." Charles II, languid with his mistress, has abandoned affairs of state for more personal affairs, with the result that the crab lice "In this Extremity they both agree,/A Commonwealth their Government shall be."[50] Another potential tyrant is brought so low that even his lice abandon him.

Thus lice provided a powerful reminder that not even the mightiest could escape the depredations of the smallest of animals—hence no one could afford to neglect his or her moral, religious, political, or personal duties. Tyranny bred corruption, and immorality resulted in bodily rot, all thanks to a parasite sucking on human flesh. Lice, whether God's instrument or nature's tool, are everywhere in premodern literature, mirroring their ubiquity on premodern bodies. Naturalists and experimenters had to justify any interest in lice, creatures that sometimes seemed to put divine creation and belief in its creator at risk. Because of the political and sexual connotations of the parasite, apparently innocent enquiry could sometimes suggest sensual excess. Robert Hooke, Margaret Cavendish, and Antonie van Leeuwenhoek all understood that there was more to a louse than an itch on the body or an object to gaze at through the microscope. Puritans and pornographers knew that lice had a message for kings and commoners. Lice, in fact, could be used as a weapon of social commentary and political satire. We will see in chapter 4 that the kind of satirical attack Dorset made on Charles II, and Alexander Wolcott directed at George III, was just one of the ways that lice could elicit laughter when twenty-first-century people would expect disgust. Entire social groups were stigmatized by being associated with lice. As with smelly bedbugs, lice bore a significance far beyond their size, carrying cultural impact on six legs.

CHAPTER FOUR

Lousy Societies

Infesting the Lower Classes and Foreigners

Jonathan Swift didn't like anyone, but he particularly loathed the Scots. Ridiculing the union of England with Scotland and Ireland, he compared Scotland to an unattractive lady who has a "natural Sluttiness; for she is always Lousy, and never without the Itch. . . . She is poor and beggarly, and gets a sorry Maintenance by pilfering wherever she comes."[1]

Scot Robert Burns had a different notion of his nation, but one of his most famous poems (translated here) describes a louse making its way across a Scotswoman:

> You ugly, creeping, blasted wonder,
> Detested, shunned by saint and sinner,
> How dare you set your foot upon her—
> Such fine a lady!
> Go somewhere else and seek your dinner
> On some poor body.
>
> Off! in some beggar's temples squat.[2]

Swift and Burns, in a few lines, set out many of the social meanings of lice: the creatures are the accomplices and companions of foreigners, women, and beggars. Even saints and sinners despise lice—whatever theological justifications they'd ever had no longer existed for these eighteenth-century wits.

Consequently, lice were the perfect weapon for satirists' black comedies or acerbic poems. In early modern England, lice stirred laughs and even shrieks of delight. They were the paramount satirical insect, used to take any target down a peg. Lice were equal opportunity ammunition, aimed not only at kings and scientists but also at social outcasts, the people whose presence most discomforted the affluent. Contemporary satirists loosed lice both at the poor and at anyone who thought himself better than the lowly.

Like bedbugs, lice crossed both the boundaries of skin and the borders of home and country. Bedbugs were loathed, but the people they infested were often portrayed as victims. Lice-ridden people or peoples, however, were considered accomplices, as deserving of disgust as any parasite.[3] There was always a moral dimension to lice commentary, either humorous or serious. Although fleas "trouble us much," pronounced the seventeenth-century natural historian Thomas Muffet, "yet they neither stink as Wall-lice [bedbugs] doe, nor is it any disgrace to a man to be troubled with them, as it is to be lowsie."[4] The historian Keith Thomas notes, "it was a disgrace to be lousy, and employment was more readily available to those who were neatly turned out."[5]

By the seventeenth and eighteenth centuries, lice were also emblems of poverty, signs of the bestiality and social degradation of their hosts. In contrast, during the Middle Ages, helping the poor was considered a good work, leading saints to embrace both the poor and their lice. But by 1500, the socially and morally fastidious distinguished between the "deserving" poor and "able-bodied" or "sturdy" beggars, able to work but refusing to do so. According to popular literature, such people bolstered their begging with fake disabilities and (very real) lice.

In an increasingly mobile and changing society, sturdy beggars represented a threat to those claiming civility and respectability—and, supposedly, freedom from lice. The disgrace of lice, and of those who had them, could threaten anyone claiming escape from the verminous stew of premodern living conditions and habits. Obsession with the down-and-outs, the homeless, and potentially dangerous social outliers was reflected in many books, ballads, and pamphlets devoted to them, perhaps because they seemed to enjoy a freedom from the social conventions confining the middle and upper classes. Upper-class individuals were expected to exercise self-control and not give into bodily demands. They were taught from childhood to feel embarrassment and shame for behaviors considered taboo in civilized society. The teenage George Washington copied out the following rule of decent behavior and courtesy: "Kill no Vermin, as Fleas, lice, ticks, etc. in the sight of others."[6] These rules were probably inspired by the Renaissance polymath Desiderius Erasmus's *The Civility of Childhood*, first published in 1530 and quickly translated into English. Writing to a young prince, Erasmus advises, "There must be neither lice nor nits. Oftentimes to scratch the head in the presence of others is a thing not very decent nor honest: as to scratch the body with nails, is a foul and filthy thing."[7]

Clearly, anyone picking lice in public was open to social ostracism and moral condemnation, not to mention guffaws—and equally clearly, no matter how much they denied it, the upper classes still had lice. Even after the well-to-do started battling lice by shaving their heads and wearing wigs, they still suffered the ignominy of the creatures' presence—making the haves as well as the have-nots targets of lice humor.

Lurking beneath the hair and clothes of early modern bodies, lice revealed the moral qualities of their hosts, uncovering filth of both body and behavior. Satire gets its bite from the moral outrage lurking beneath the joke. Both the morality and pretentiousness of Robert Burns's Scottish lady are undercut by the louse meandering around her bonnet, breaking the boundaries of class and gender. The poet is at first appalled that the louse has dared to broach a fair miss's headpiece, rather than a "an old wife's flannel cap" or the undervest of "some small ragged boy," but ultimately he names the lady as the cause of the lousy incursion: "O Jenny," he chides her, "do not toss your head,/And set your beauties all abroad!/You little know what cursed speed/The blastie's making!"[8]

Burns's litany of those deserving of the louse's presence—poor old women, ragged boys, beggars, and young misses—demonstrates the range and potency of lousiness. One mark of social superiority was to designate the other, to whom one is superior. When that other bore the physical mark of otherness—the louse or its bite—a new and powerful social designator had been identified.

The louse's accomplice in social branding was hair—the human head louse was a specialized creature, needing hair to live. Hair, along with skin, is the most malleable of human characteristics, alterable to change the social significance of its possessor. Different classes wear different hairstyles; different nations shave, chop, and color their hair in their own ways; and different haircuts distinguish various stages of human growth and signify different identities. Hair is a ready instrument to express social conformity or alienation.[9] Hair accordingly gives not only a home but also meaning to lice, and lice in turn give meaning to hair. An exploration of the social significance of lice becomes thick and hirsute—Burns's elegy to the louse ends with the famous lines, "O would some Power the gift to give us/To see ourselves as others see us!"[10] For the poet, all people are ridden with self-delusion, revealed by the louse in its promiscuous wanderings. Men and women may change their hairstyles, but the audacious louse undermines their pretensions of cultural and aesthetic superiority. Hence it became ever more necessary to insist that these insect parasites are natural only to the underclasses, not to their well-coiffed betters.

Lice lived on beggars and thieves, and characterized any group the early modern English considered alien—including the neighboring Scots and Irish. Given the strong connection between lice and disgust, it was easy for the English to extend their antipathy for the lower classes to the inhabitants of other countries and continents. These peoples were "uncivilized," according to their conquerors, because they were bestial—and instead of cultivating their manners and restraining their animalistic urges, they took no exception to having lice and even ate them. What greater proof could there be of the superiority of English civilization?

Lice therefore offered a gauge of the level of disdain directed at various targets in early modern times—particularly prisoners and paupers, social misfits, and foreigners. Sometimes even ridicule became too tame, and the lice ridden were simply condemned as revolting and vicious. Political animosity and social antipathies look particularly raw through the prism of lice.

Satirical Lice

Early modern satirists and moralists often took a dim view of humanity as a whole, a scorn expressed in lice allusions. In the 1637 *Tale of a Tub*, the dramatist Ben Jonson captured the lice–human relationship in the sneer, "I care not, I, not three skips of a Lowse for you." As Muffet would later say in his discussion of lice, "we have an English Proverb of a poor man, *He is not worth a Lowse.*"[11]

Turning this sentiment upside down could puncture pretension at all social levels: sometimes a man is not worth a louse because the louse is better than the man. The Dutch humanist Daniel Heinsius, who praised the "fricative" pleasure derived from scratching a louse bite, devoted a whole oration to lice's wonderful qualities. *In Praise of the Louse* (*Laus Pediculi*), published in 1635, features a lawyer defending a louse before the "Worshipfull Masters and Wardens of Beggars Hall"—an audience clearly well acquainted with the defendant. He praises the louse, man's "ever trusty companion" that "suffers under the tyrannicall oppression of men, and is made by them as contemptibly infamous as they can." Indeed, in many ways, lice are better than men because "Man . . . is borne of stone, but the Lice are borne of Man. So much the nobler in his originall, as a man is nobler then a stone." This creature that dwells on the most rational part of man, Heinsius tells us, benefits from its good neighborhood and possesses understanding, prudence, and wisdom. In fact, the louse is most "busied in husbandry and domestique affaires, [and] all the spare time remaining from the exercise and care of feeding, it bestoweth on contemplation, and rest." These contemplative creatures are therefore almost Pythagorean in their silence while still being as sociable as Aristotle desired men to be, living and eating with their fellows and their human hosts. And unlike other beings, the louse does not abandon one in adversity but "is a true companion and attendant to poverty," clinging to his host even when a man is in chains or on the gallows.[12]

Heinsius, contrasting the fidelity of lice with the faithlessness of people, was not alone in hurling the louse as a weapon against human pretentiousness. The poet Robert Heath (1575–1649), clearly inspired by the *Laus Pediculi*, praised the louse, urging men to "Observe his [the louse's] generous disposition in his Sedate constancy of Affection, scorning to leave his friend in his worst of fortunes, but will faithfully accompany him from the Court even to the Camp or Prison. . . . next to Man, the Louse is the Noblest Creature."[13]

In these descriptions, it seems that lice possess all the qualities of a good man, a litany of the aspirational traits of early modern Englishmen: courage in adversity, prudence in daily affairs, loyalty in all circumstances, sociability with their hosts and fellows, contemplative in times of rest. Lice keep faithful company with individuals of every social class, but especially those who fall into poverty or crime, who perhaps particularly need to remember the virtues possessed by the insect. On the wheel of fortune, a valued early modern image, the louse is the metaphoric linchpin, reminding kings that they can become beggars and beggars that they can become kings.

A number of early modern satirical works exploited this fear of boundary-jumping lice by following their wanderings up and down the social hierarchy. In one eighteenth-century tale, a louse proclaims that he was "Got in an Alley near St. Bow, born on a sturdy Beggar's Smock" but deserted his birthplace to lodge on a lawyer—attitudes toward some professions are timeless—declaring, "I with wond'rous Art did bite / The Man who cozen'd all Mankind." From the lawyer, the louse visited a judge, then climbed aboard a lady, and from her to a statesman to a coquette, and down the social scale to a servant and then a whore. The moral of the story is that, "So after all my Care and Strife, Pleas'd like mankind, like him in Pain, Both he and I must yield up Life, and quiet, turn to Dust again." And so the louse is "Exempt from ev'ry Fear, But the untimely, deadly Crack."[14]

Just a few years after that louse recalled his peregrinations, another sententious insect told a similar story. In the biweekly "Adventurer," a louse sermonizes in a dream to the possessor of the head he now inhabits, teaching that "Life is a state of perpetual peril and inquietude." Although the louse "does not remember that I have brought calamity upon myself from any uncanny deviations from either virtue or prudence," his wife and children were "crushed to atoms" when the head of the charity boy on which they lived was washed, and he himself was brushed by the boy into a basket of laundry. The louse travels from the linen to the neck of a "celebrated toast," and from her to a "battered beau," who nearly kills him with his primping at court. The louse then leaves the courtier for his valet, and after the servant is sacked for his new master, a barber. Further adventures drop him onto the head of an experimental philosopher—perhaps the louse was reading Hooke's *Micrographia*—and then to a doctor whose nurse almost includes him in an antidote for jaundice (a spoonful of lice in milk), which she administers to a six-year-old boy. The barber shaves the boy, and the louse decamps for his shaving cloth and then to the writer of the essay. The tribulations of the louse cause the writer "to burst into a fit of immoderate laughter," which is followed by the louse's reflection that "The life of man is no less exposed to evil; and that all his expectations of security and happiness in temporal possessions, are equally chimerical and absurd."[15]

But lice did not always teach moral lessons. More often in satires, lice are identified with the loathsomeness of their hosts, a manifestation of the humans' evil and base nature, as well as their sweat and bodily humors. In a fanciful pamphlet, the Jacobean dramatist Thomas Dekker (1572–1635) describes a meeting of the beggars of England: "They are the idle drones of a Countrie, the Caterpillers of a common wealth, and the Aegiptian lice of a Kingdome." His evoking Egypt recalls both the scriptural role of lice and their association with Gypsies—then thought to come from Egypt—and paupers, whose effect on the body politic reflect lice's impact on the human body. On this occasion, the beggars wear "hansome cleane linnen" unlike their usual lousy attire and are organized into ranks "according to degrees of Superioritie and Inferioritie in our Societie." Ranked in a hierarchy like all classes of society in early modern England, beggars compose a corporate group, a "colledge." They meet regularly in a "great Hall," which on this occasion was "so full it swarmed with them," a verb verminous in itself. The beggars "are a people for whom the world cares not, neither care they for the world: they are freemen, yet scorne to live in Cities: great travellers they are and yet never from home, poore they are, and yet have their dyet from the best mens tables."[16] Like lice, beggars live off others, benefitting from the society they scorn, and which scorns them in turn. The attitude, and even the image, might be familiar from our own politics.

Another satire, written by Edward Ward (1667–1731), a participant in the community of hack writers and impoverished journalists referred to as "Grub Street" in the eighteenth century, describes a club of beggars: "This Society of Old Bearded Hypocrites, Wooden leg'd Implorers of good Christian Charity, strolling Clapperdudgeons, lymping Dissemblers, sham disabled Seamen, Blind Gunpowder-blasted Mumpers, and old broken limb'd Labourers" who entertain themselves by watching one of their fellows who "fell to fingering his Collar, conveying his little Foes that he happens to take Prisoner between Finger and Thumb, from his Neck to his Mouth, that he may bite the Biters which he dispatches so naturally, that it is hard to distinguish whether he is in Jest or in Earnest: Thus he recreates himself, and diverts the Company, who cannot forbear shrugging at the lousy Performance, as if they itch'd by Sympathy."[17]

The beggar "bites the biters" just as the beggars put on a verminous performance with fake disabilities and fraudulent social credentials, preying on the credulity and kindness of the society they despise. Similarly, criminals attack their social betters—and are also well acquainted with lice. In jail, a prisoner can expect the company of the parasite, even if he doesn't quite share Heinsius's appreciation for their fidelity. In an early seventeenth-century comedy, an imprisoned character refuses to give up his claim to marry a lady of substance: "First I will stinck in Jayle, be eaten with Lyce, Endure an object worse then the Devill himself, And that's ten Sergeants peeping through the grates upon my lowsie linen."[18] The seventeenth-century humor may lose something in the twenty-first century.

Newgate Prison was particularly notorious for its vermin. In one of William Hogarth's most famous series of engravings, *Industry and Idleness* (1754), "The Idle 'Prentice" begins his downward spiral to Newgate and the gallows by watching an ignoble trio gamble on a coffin and scratching his head, while the righteous attend church in the background. The nineteenth-century editor of Hogarth's works comments, "The hand of the boy, employed upon his head, and that of the shoe-black, in his bosom, are expressive of filth and vermin; and show that our hero is within a step of being overspread with the beggarly contagion."[19]

Prisoners and beggars are plagued with lice, but in the literature of early modern Europe, the lower classes are sometimes vermin's accomplices rather than their victims. Both French and English works describe the down-and-outs weaponizing their creepy companions, just as eighteenth-century Londoners feared that their servants either knowingly or inadvertently contaminated their betters with bedbugs. In the sixteenth century, Francois Rabelais

William Hogarth, *Industry and Idleness, Plate 3: The Idle Prentice at Play in the Church-yard during Divine Service*, engraving from *The Works of William Hogarth*, 1833. Project Gutenberg, https://www.gutenberg.org/files/22500/22500-h/22500-h.htm

(who knew something about satire) describes the practices of Panurge, a miscreant friend of the giant Pantagruel: "In another [pocket] he had a great many little hornes full of fleas and lice, which he borrowed from the beggars of St. *Innocent*, and cast them with small canes or quills to write with into the necks of the daintiest Gentlewomen that he could finde, yea even in the Church."[20] A few years later, the English divine and historian Thomas Fuller (1608–61) reported on "Beggars, who breed Vermine in their own bodies, and then blow them on the cloaths of others."[21] In a late seventeenth-century story attributed to the Earl of Rochester, a pickpocket seeking to steal a watch comes up with a stratagem:

> I had a Quill of living Lice in my Pocket, prepared for such a design, this I took
> out, and unstopping it at both ends, with a gentle blast fixed them upon the
> Back and Shoulders of my Spark: the six-footed animals no sooner found

themselves at liberty, but they began to crawl . . . so that they were soon perceived by the by-standers, some laughed, and others told him, he was well Guarded.

This diversion results in success for the thief, as his mark, attempting to knock the vermin off, is more than sufficiently distracted.[22]

Perhaps the most astonishing association of lice and weapons, likely inspired by Rabelais, occurs in Victor Hugo's *Les Misérables*. Jean Valjean and Cosette observe a procession of prisoners chained together in an almost endless line, wearing wretched rags barely covering their red and unhealthy skin (*des dartres et des rougeurs malsines*)—perhaps the legacy of body lice. The convicts, angered by the jeering crowd around them, "had quill pipes through which they blew vermin at the crowd, singling out the women" (*avaient à la bouche des tuyaux de plume d'où ils soufflaient de la vermine sur la foule, choisissant les femmes*). *Les Misérables* was published in 1862 and translated into English the next year. In reclaiming agency for the convicts, Hugo implicitly condemns the bourgeoisie who watch them with expressions of "moronic bliss"—in some ways the spectators are, if in a different sense, as lousy as the convicts. When Cosette asks Jean Valjean, "Father, are these men?" the former convict, subdued by the memory of his own captivity, responds, "At times."[23]

Hair, Lice, and Civilization

We saw in chapter 3 how lice could be used in political and religious diatribes, employed to attack the foibles and follicles of kings or the hirsute habits of political and religious adversaries. The presence of lice in pubic hair enabled both the participants in sexual dalliances and their critics to link pornography and women with vermin. Hair is a weighty cultural commodity, and those claiming social preeminence wanted to dislodge the vermin that inhabited their coiffures and wigs, something not easily done before the arrival of modern hygiene.

The shame generated by lice in one's hair was something new and modern, demonstrating how much the upper classes wanted to control their bodies. By the late seventeenth century, hygiene became both a private occupation and a defense against social disgrace. Before then, communal delousing was a common practice. People sat around picking lice off each other. The image on page 92, from a late fifteenth-century health manual, shows an upper-class woman delousing a man. Both seem pretty happy about it.

Bedbugs, delousing, from *Hortus Sanitatis*, 1536. Courtesy of the Wellcome Collection, https://wellcomecollection.org/works/pt8cgg6b

The French historian Emmanuel Le Roy Ladurie describes familial delousing in a fourteenth-century French village: "Benete and Alazais Rives were being deloused in the sun by their daughters."[24] And the fifteenth-century Englishwoman Margery Kempe (ca. 1373 to ca. 1438) told of a group of poor people she met in Germany while on pilgrimage to Jerusalem: "her companions took off their clothes, and, sitting around naked, picked themselves for vermin. This creature [Margery] was afraid to take off her clothes as her fellows did, and therefore, through mixing with them was dreadfully bitten and stung both day and night."[25]

Margery may have been particularly afraid of lice because women, perhaps because of their long hair, were considered especially prone to the vermin. The medical understanding of hair in premodern times usually related it to the theory of the four humors, whose balance or imbalance was thought to produce health or ill health. In the eighteenth century, the theory reinforced the rising tide of civility and hygiene. *Aristotle's New Book of Problems*, published anonymously in 1725, maintained that hair is "an Excrement generated and form'd of the most gross and earthly Superfluities of the third

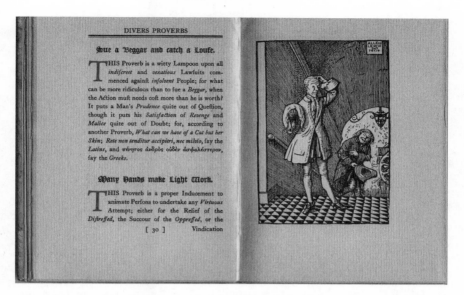

"Sue a Beggar and catch a Louse" and "Many Hands make Light Work," woodcuts from Nathan Bailey's *Dictionary of Proverbs*, 1917. Courtesy of Liam R. Quin, https://www.fromoldbooks.org/proverbs/pages/pp30-31/

Concoction; The Benefit of it is, that consuming the gross, fuliginous, and sooty Excrements of the Brain, it becomes a Cover and Ornament to the Head, and finding there its most proper matter for encrease, it grows very long, especially in Women, who have moister Brains than Men."[26] Moist brains breed lice, as Dr. Culpeper explained in chapter 3, but even men could become prey to the bloodsucking insect. The danger of lice to members of the middling and upper classes is depicted in this eighteenth-century woodcut.

The above proverb, going back to the seventeenth century, is meant as advice to the legal profession not to go after the indigent, but it might also warn of the ease with which lice can pass from beggar to gentleman, cutting right through the social order. The lawyer in the engraving holds a wig—a symbol of his status—as he scratches his shaven head, revealing that lice are no respecters of class. Wigs and hair reveal how much new ideals of civility had taken hold by the late seventeenth century, but they also reveal anxiety about social class, and the engraving shows how little protection wigs could provide.

In 1662, as seen in chapter 1, the famous diarist Samuel Pepys could laugh at lice. But later in the decade, the creatures left him less lighthearted. Pepys had constant trouble in obtaining a vermin-free wig; on one occasion in 1664, he went to his barber "to have my Periwigg he lately made me cleansed of its nits, which vexed me cruelly that he should put such a thing into my hands."[27] Still, the prospect of infestation—Pepys also worried that the hair used in wig making could carry the plague—did not prevent him or many others from adopting the fashion, which became prevalent in all classes of society in the late seventeenth and eighteenth centuries. Since it was not yet common to wash hair with water—which many at the time considered dangerous, not always wrongly—wigs freed their owners from needing to powder their own greasy hair.[28]

Pepys may have been amused by his lousy accommodations while traveling, but his forbearance ran out when the parasite appeared in his own home. In 1663, he berated his wife for the house's lack of cleanliness, calling her a "beggar," and she retaliated by calling him a "pricklouse, which vexed me."[29] Both husband and wife were invoking lice's low-class aura against each other, and the insults were calculated to sting: Elizabeth Pepys, if not a beggar, came from a lower-middle-class family, and Pepys himself was the son of a tailor. "Pricklouse" was a slang term for a tailor in the seventeenth century, referring either to tailors' remaking lice-ridden clothes or to the prick of the needle resembling the bite of the insect. Tailors were socially inferior to other craft guilds and all merchant guilds.[30] Thus Pepys's wife was reminding him of his own ignoble, not to say verminous, origins, as well as his lack of cleanliness.

Pepys became Clerk of the Acts of the Navy Board in 1661 and was extremely sensitive about his roots, a sensitivity reflected in his constant efforts to create an ordered—and presumably lice-free—environment in his home.[31] He lamented that a serving girl had run away after "being cleansed this day of lice by my wife, and good, new clothes put on her." And he was himself humiliated when he was "mightily troubled with an itching all over my body," which he took to be the result of lice bites: "I found this afternoon that all my body is inflamed, and my face in a sad redness and swelling and pimpled, so that I was . . . not only sick but ashamed of my self to see myself so changed in countenance."[32] His exasperation came to a head, so to speak, in 1669, with the following incident:

> So to my wife's chamber, and there supped, and got her [to] cut my hair and
> look [at] my shirt, for I have itched mightily these 6 or 7 days, and when all
> comes to all she finds that I am lousy, having found in my head and body about

twenty lice, little and great, which I wonder at, being more than I have had I believe these 20 years. . . . So how they come I know not, but presently did shift myself, and so shall be rid of them, and cut my hair close to my head, and so with much content to bed.[33]

Clearly, the civilizing process had done its work on Samuel Pepys. If the revulsion caused by bedbugs was a sign of modernity, the shame and embarrassment caused by lice showed the surfacing of bourgeois respectability, a sensibility that turned morality into manners. One could argue that the vision of the proper English gentleman (or at least one proper Englishman)—well educated, well spoken, and newly clean (notably of lice)—emerged between 1662, when Pepys was amused by his lousy bedding, and 1669, when he went to bed deloused and content.

Pepys's exchanges with his wigmaker show that the new standards of civility generated a new profession. Just as exterminators would later appear, applying their expertise to bug catching, professional hairdressers and wigmakers started to advertise their services at the end of the seventeenth century. By the end of the next century, reported one contemporary, England had twenty thousand hairdressers. The number may be exaggerated, but a 1795 tax on hair powder brought in £210,136. (Unfortunately for the Crown's revenue, after the French Revolution, the fashion of powdering wigs—and the tax on powder—disappeared.[34])

During the eighteenth century, men of all classes wore powdered wigs, and upper-class women sported really big hair—towers of hair. "For those of the higher orders," recounts the historian Don Herzog, "hairdressing was an elaborate business. Hair was plastered, powdered (up to two pounds of powder per head), curled, and lubricated with pomatum or bear grease or Macassar oil. This mass of stuff had to be combed out and reapplied daily—it must have gotten horribly messy while sleeping, and anyway it must have supported an imposing population of flora and fauna—which made for lively demand for hairdressers, the more expert the better."[35]

Fashion demanded sacrifice. Through the molds and hair pieces used to produce the towering effect, big hair reopened the gates to the verminous enemy that the new standards were seeking so hard to expel from the body.

On page 96, the 1779 portrait of Georgiana Cavendish, Duchess of Devonshire, captures the heights to which these hairstyles rose. The portraitist Lady Diana Beauclerk (1734–1808) surely knew the dangers inherent in tall hair—and not from just falling over. She was married to the infamous Topham

Lady Diana Beauclerk, *Lady Georgiana Cavendish, Duchess of Devonshire*, stipple engraving, 1779. Wikipedia, https://en.wikipedia.org/wiki/File:Stipple_engraving_of _Georgiana_Devonshire_after_Diana_Beauclerk.jpg

Beauclerk, a man renowned for his filthiness. One story about him brings together many of the patterns found in the history of lice:

> The elegant and accomplished gentleman . . . was . . . what the French call
> *cynique* in his personal habits beyond what one would have thought possible in
> anyone but a beggar or a gypsy. He and Lady Di made part of a great Christmas
> party at Blenheim, where soon after the company were all met, they all found
> themselves as strangely annoyed as the Court of Pharaoh were of old by certain

visitants—"in all their quarters"—It was in the days of powder and pomatum, when stiff frizzling and curling, with hot irons and black pins, made the entrance of combs extremely difficult—in short, the distress became unspeakable.[36]

But despite multiple reports of vermin in upper-class English coiffures—only their hairdressers knew for sure, and they weren't talking—the English nevertheless condemned other peoples for their lack of hygiene and failure to comb their hair. This condemnation incorporated not only ethnic and nationalistic strains but also religious prejudices and judgments about who was and wasn't civilized. Pepys notes that on another walk, "drinking my morning draft of whay, by the way, to York House, where the Russia Embassador do lie; and there I saw his people go up and down louseing themselves."[37] For Pepys, delousing was necessarily private to avoid the shame associated with vermin; the public personal cleansing by the members of the Russian mission clearly indicated that they were not yet part of civilized society. Some forty years later, the czar Peter the Great would agree. As part of his effort to Westernize Russia, he ordered his nobles to shave off their beards or pay a hefty fine—presumably razing some densely populated verminous habitats.

But the lash of lousiness could still be directed against the English themselves. In Jonathan Swift's *Gulliver's Travels*, the king of the giant Brobdingnagians imagines Europeans, whom Gulliver had described to the monarch as "the most pernicious race of little odious vermin that nature ever suffered to crawl upon the surface of the earth." But the giant foreigners, used by Swift to mock his own countrymen, have their own species of lice. Gulliver relates, "the most hateful Sight of all was the Lice crawling on their Cloaths. I could see distinctly the Limbs of these Vermin with my naked Eye, much better than those of an European Louse through a Microscope, and their Snouts with which they rooted like Swine."[38] In the mythical geography of *Gulliver's Travels*, the country of Brobdingnag is a peninsula off the northwestern coast of North America. Apparently, lice were a universal phenomenon that the English would find wherever they went.

Foreign Lice and the Lousy Other

If hair itself signified difference, not surprisingly, the perceived relationship between foreigners or natives of new places and the insects inhabiting their hair and bodies also justified English and American aggression and expansion. Other societies could, and should, be exploited because they were bestial, like apes who picked lice out of their hair. From early on, the English characterized the subjugated Irish, Scots, and Africans as bug ridden. Thomas Muffet, managing to disparage both Irish and Africans, believed "that the Blackmoors [of St. Thomas] there are full of Lice, but the white men are free of that trouble: All Ireland is noted for this, that it swarms almost with Lice."[39] Edmund Spenser shared a similar attitude toward the Irish, reflecting his time there as Lord Lieutenant; after his description of the mantles worn by dissolute Irish women, he adds, "And as for all other good women which love to doe but little worke, how handsome it is to lye in and sleepe, or to louse themselves in the Sun-shine."[40]

Another sixteenth-century author repeated the truism that Ireland was free of snakes but noted, "the men are not free from lice, which cometh of sluttish and filthy use."[41] The English were clearly refining the moral judgments that would shape their later views of other foreigners: they are dirty, sexually promiscuous, and lazy, and share many of the qualities of beggars—including taking pleasure in delousing themselves. By the seventeenth century, one account suggests, the English simply concluded the Irish were themselves lice. Oliver Cromwell is often credited with the saying, "nits make Lice," which he supposedly used at the Battle of Drogheda in 1649 to justify the killing of women and children as well as rebellious Irish warriors. Modern Cromwell fans deny that he was ever so heartless and genocidal, but a mid-seventeenth-century English poet said of one of Cromwell's commanders in the massacre of Irish Catholics: "brave Sir Charles Coote . . . by (by good advice), Did kill the Nitts, that they might not growe Lice."[42]

Similar, if not quite so deadly, assumptions characterize English attitudes toward the Scots. In beggars' cant, the language of beggars and thieves, Scotland was referred to as "louse land." Since beggars themselves were considered the epitome of lousiness, it seems even the lowest of English society cherished a verminous view of their northern neighbors. As one early eighteenth-century wit recounted, "Unkle Dering of mine was wont to say, that he had been a fortnight in Scotland, and yet had got their present State at his fingers end. I was not so afraid of being lousie. Since tis well known, that set a louse upon a table, and he shall dutifully direct his course Northward towards his Mother Country. So I was sure if I caught any, to leave 'em behind me."[43]

Just as English travelers to Ireland and Scotland expected to find disgusting habits and practices among their Celtic neighbors, they also anticipated them when journeying farther afield. Travelers' tales, going back to Odysseus in antiquity and Marco Polo and John Mandeville in the Middle Ages, are filled with wondrous monsters and exotic human beings. Encounters with the other, wherever the other might be found, both affirmed the superiority of the English traveler (and reader) and provided a source of horrified entertainment. The strange and distant—the exotic—enthralled the homebound. In the many travelers' accounts of their experiences, featuring varying degrees of accuracy and fantasy, lice frequently crept in—and as one would expect, not in a way to compliment the locals.

Not surprisingly, accounts of the disgusting habits of lice-bearing peoples frequently focused on their hair, often the most obvious marker of cultural differences. Eastern Europeans were believed to be subject to a particularly disgusting, lice-inflected display called the *plica polonica*, or Polish plait. A plica was a long lock of hair growing in one or two clumps that, according to eastern European folkloric beliefs, had to be left unwashed or uncut to protect the body from disease. If it were cut, blood would swell from the spot, and swellings and fever would result.

Westerners reversed the etiology and believed that the plica itself caused serious illness.[44] According to a seventeenth-century text, "The Russians and the Cossacks are afflicted with a disease called by the Physitians, *Plica*, and in the language of the Countrey *Goschest*, [when] they are seized with it, loose the use of their Limbs, as Paralitical persons doe, feeling great pains in their Nerves."[45] Unlike phthiriasis, the lousy disease, this was a real condition, which modern epidemiologists say develops "as a result of an immune response of the human body to head lice bites."[46] In early modern times, the English observers of this disease knew that it was associated with lice but thought that the parasites were the result rather than the cause of the problem. Thus the Scottish physician Andrew Duncan (1773–1832) described it as "a disease endemic in Poland, and the neighbouring countries, in which a morbid matter is critically deposited upon the hair, and binds it together in such a manner, that it becomes impossible to unravel it." He continued, "If the whole hair be affected, it forms a kind of cap; if it flows out only from single hairs, then several rope-like plicas are produced. Some days after it has been formed, it begins to smell something like rancid fat; and upon being touched, causes a prickling, unpleasant sensation in the fingers. At the same time, lice suddenly appear in such disgusting numbers, that the patient suffers more from them than from the disease."[47]

Most commentators thought that the Russians and Poles had acquired the custom of growing plaits from the Tartars, viewed as an even more disgusting society. The geographer Michael Adams argued, using anthropological rather than medical criteria,

> Some of the tribes [of Tartary] are far more filthy than others, and in this particular none exceed the Kamtschatkans, who are said never to wash their hands or face, nor cut their nails . . . Both men and women plait their hair in two locks, binding the ends with small cords. If any hair happens to start up, they sew it down with thread, to make it lie close, which means their heads swarm with vermin, which they scrape off with their hands.[48]

Thus this eighteenth-century version of dreadlocks invoked the same kind of condemnations that some observers feel for the current hairstyle. Even the *Philosophical Transactions of the Royal Society* condemned it as a condition produced by "nastiness, from not combing their hair," and a German doctor writing to the society observed in 1724:

> The great number of people in Poland, who are troubled with this plica, first made me reflect, whether it was a real distemper or not? But I am now convinced, that their swinest way of living, and the common opinion of the people, namely, that this lock of hair cannot be taken off without danger of their lives, have contributed more to this complaint than any real indisposition of the body; considering that it is the middling or poorer sort of people who are troubled with it, who one cannot see without horror. But no German, of who great numbers live in that country, ever has such a thing grow.[49]

Perhaps no German was ever afflicted with this condition, but it was not completely unknown in England, where the matted hair was referred to as an elflock. *Romeo and Juliet* cites the style during a speech by Mercutio full of insect allusions:

> This is that very Mab
> That plats the manes of horses in the night,
> And bakes the elflocks in foul sluttish hairs,
> Which once untangled, much misfortune bodes. (1.4.93–6)

Mab was Queen of the Fairies, so this condition had magical and perhaps devilish connotations for the English. Thomas Hall (1610–65), a Puritan preacher, included a reference to the Polish plica in his condemnation of long hair, where he insists that this loathsome and un-Christian disease now afflicts the English as well as the Poles:

Have you not been inform' o th' hand
Of God on Poland lately laid;
Enough to make all Lands afraid,
And your long dangles stand on end?
Feare him that did that Plica send,
And those sad Crawlers: and hath more
Unheard of Judgements still in store,
Than the vast Heaven hath glorious stars,
Or those your delicate Heads have hairs.[50]

The plica did not appear among the long-haired English, however, what-ever Puritans might hope, and by the 1700s it was associated only with eastern Europeans, especially the Jews of Poland and Russia. The traveler Andrew Duncan wrote in 1796, "The Jews, the most bigoted of all beings, have their own customs and cures: they never permit it to be cut off; and nothing can be more disgusting, than one whose beard and whiskers are affected with this disease."[51]

The disgust that Jews and Poles generated in Anglo-Saxon writers would be echoed in subsequent centuries by racists who associated inferior peoples with vermin. By the twentieth century, Jews were often linked with lice, until the insecticide developed during the First World War to destroy lice in the trenches was employed by the Nazis to kill Jews in the gas chambers. Al-though most historians of anti-Semitism distinguish between the religious prejudices of earlier Europe and the racial attitudes of modern times, the attitudes about lice and Jews, and lice and Poles, and lice and Tartars reveal a direct line of verminous vitriol directed at the other. It is, after all, the Ger-mans living in Poland who escape lice.

But at least eastern Europeans and Jews were never accused of eating their lice, a common theme in accounts of exotic non-Europeans. The Spanish historian Peter Martyr d'Anghiera (1452–1526) may have begun the legend of lice-eating natives. In his description of Latin America, he wrote, "When these Indians are infected with this filthiness, they dresse and cleanse one another. And they that exercise this, are for the most part women, who eat all they take."[52] Similarly, the French Jesuit explorer Louis Hennepin (1626–1705) recounted that when exploring what is now the northern United States, "I was also much surpriz'd one day to see an Old decrepit Woman, who was employ'd in biting a Child's hair and devouring the Lice that were in it."[53] In an account of an expedition seeking the Northwest Passage in northern Canada, the explorer and naturalist Samuel Hearne (1745–92) tells a similar tale of Eskimos:

> Their clothing, which chiefly consists of deer skins . . . makes them very sub-
> ject to be lousy; but that is so far from being thought a disgrace, that the best
> among them amuse themselves with catching and eating these vermin. . . . My
> old guide, Matonabbee, was so remarkably fond of those little vermin, that he
> frequently set five or six of his strapping wives to work to louse their hairy
> deer-skins, the produce of which being always very considerable, he eagerly
> received with both hands, and licked them in as fast, and with as good a grace,
> as any European epicure would the mites in a cheese. . . . I had no inclination
> to accustom myself to such dainties.[54]

According to travel accounts, European lice avoided the fate of their foreign brethren by deserting their hosts when the voyagers passed the equa-
tor. The Spanish explorer Gonzalo Fernandez de Oviedo y Valdés (1478–1557)
mentions that during a journey along the coast of Latin America, the lice on
soldiers "died and forsook them, [but] sodainly in their repassing by the same
clime (as though these Lice had tarried for them in that place) they can by no
meanes avoide them for the space of certaine daies, although they change
their shirts two or three times in a day."[55] This legend found its way into literature when Don Quixote instructs Sancho about the fate of European lice:

> You shall understand *Sancho,* that when the Spanyards, and those that im-
> barque themselves at *Cadiz,* to goe to the *East Indies,* one of the greatest signes
> they have, to know whether they have passed the *aequinoctial,* is, that all men
> that are in the Ship, their Lice die upon them, and not one remaines with them,
> not in the Vessel, though they would give their weight in gold for him: so that
> *Sancho,* thou maist put thy hand to thy thigh, and if thou meet with any live
> thing, we shall be out of doubt; if thou findest nothing, then we have passed
> the Line.[56]

The Spanish weren't the only ones to experience the peregrinations of
lice. In the seventeenth century, the Dutch and the English began a process
of seizing and exploiting the lands of tribes they called the "Hottentots" in
South Africa. The Hottentots, according to one eighteenth-century source,
"swarm with lice," although "none will live on the body of a European at the
cape." Moreover, "whatever place one of these insects alight on, is deemed
sacred by the Hottentots." He declared that "the Hottentots are certainly the
lousiest people in the world." They, of course, "devour" lice out of a sense of

revenge. And they are not ashamed about delousing themselves in public "but pursue the Game, let who will appear before 'em, with as much Countenance as we do the most laudable Employments or Diversions."[57] Once again, the bestiality of the foreigner is indicated by the presence of lice on their bodies and their menu, while lice avoid civilized people.

But as John Southall found out when he visited Jamaica and learned the secret of his bedbug-killing elixir, sometimes the natives knew more than their European betters. When another supposed traveler in Africa, after spending the night in bed with sand lice, found himself covered with red and purple spots that itched terribly, he showed "my old host the marks they had left upon me, which made him laugh; but he gave me some elephant's fat, and told me I must anoint myself." This travel account was actually fraudulent, but its illustration of European and English attitudes toward Africans, and lice, was unquestionable.[58]

As the so-called civilized British and American societies became increasingly concerned with manners and appearance, they could not tolerate association with their one-time personal guest, the louse. Their dismissal of outsider groups within their own societies as lousy expanded to cover entire nations, groups, and continents. Like the Irish, the Scots, and English beggars, the alien peoples were without appropriate shame and deserved the condemnation of the civilized. Their appearance and appetites testified to their bestiality and otherness. Although lice had been a way to satirize the pretentious and teach a lesson about the fragility of borders in any society, increasingly they became a signifier of immorality, inferiority, and otherness. In the nineteenth century, lice figure prominently in the emerging debate about racial types, even in the scientific musings of Charles Darwin. By the twentieth century, the presence of lice made the battlefields of Europe particularly loathsome. And by the time of the Second World War, lice played an even more prominent role in the genocidal imagination of the Nazis—"nits become lice" evolved into the belief that Jews, Gypsies, and the mentally handicapped were simply forms of lice that required extermination.

CHAPTER FIVE

The Perils of Lice in the Modern World

In 1834, Charles Darwin had heard a lot about lice, saying, "These disgusting vermin are very abundant in Chiloe: several people have assured me that they are quite different from the Lice in England: they are said to be much larger and softer (hence will not crack under the nail) they infest the body even more than the head."[1] Moreover, he recounted that an English surgeon on a whaling ship had told him that lice-infested natives from the Sandwich Islands could not free themselves from their invaders, "which were blacker and different from these, or any lice, which he ever saw." But if the lice tried to infest the English sailors on board, the insects died in three or four days. Apparently, these foreign lice had a particular palate—Europeans would not do. "If these facts were verified their interest would be great," mused Darwin. "Man springing from one stock according [to] his varieties having different species of parasites."[2]

In fact, some nineteenth-century naturalists and anthropologists thought that the varieties among lice might indeed mean that man, rather than evolving from a single origin (monogenesis), was instead the product of pluralism (polygenesis), a natural or divinely ordered series of creative acts producing different human races.[3] Many, including some scientists, used polygenesis to argue for differences among racial types, that Caucasians were better than "Mongoloids" and "Negroids"—a position that could justify imperialism and slavery, invoking entomology in support of superiority.[4]

In the nineteenth and twentieth centuries, such judgments wove race and lice into a complex and sordid history, encompassing the annihilation of millions of Jews, homosexuals, and the mentally challenged. From its sixteenth- and seventeenth-century role as a satire on human presumption, the cultural meaning of lice took a more sinister turn. The state of being lousy, laughable to a seventeenth-century satirist like Heinsius or annoying to the increasingly urbane and sophisticated Samuel Pepys, became by the nineteenth century evidence of not just social but also racial inferiority. Disgust turned into dehumanization. As the so-called civilized British gentleman and lady became increasingly concerned about manners and appearance, they could tolerate no association with their one-time familiar, the louse, dismissing entire nations, groups, and continents as inferior, as, in fact, lousy.

Not surprisingly, the English and Americans linked lice to the Irish and to African slaves. Although not yet understanding that lice spread typhus, the genteel nevertheless connected sickness with the parasite. Typhus drove mortality rates during the Irish Potato Famine and permeated jails and prisons, and indeed anyplace where clean clothes were an impossible dream. Thousands of Irish died of typhus in so-called coffin ships awaiting admission to Canada in 1848. An 1852 article in *Punch* compared the Irish with West Indian "Negroes" and warned that they would "crawl like wingless vermin over the country."[5]

The English considered the Irish, and indeed any foreigners, to be dirty and lice ridden, and thus savage. Cleanliness, over the nineteenth century, became an essential element of civilization. Puritan and Methodist teachings, public policy, private enterprise, and domestic economy all saw hygiene as partnered with virtue—and nothing signified filth as much as lice. The subject of lice became so taboo that the entomologist Alpheus Spring Packard (1839–1905) noted in 1870, "the creature itself has been banished from the society of the good and respectable." But Packard admitted, making a moral as well as a social observation, "Then have we not in the very centres of civilization the poor and degraded, which are most faithfully attended by these revolting satellites!"[6]

Darwin saw the same connection in "idiots," or children with microcephaly: "One idiot is described as often using his mouth in aid of his hands, whilst hunting for lice. They are often filthy in their habits, and have no sense of decency; and several cases have been published of their bodies being remarkably hairy."[7]

Moral judgment fueled many popular accounts of the lice ridden. Explaining pubic lice, one American doctor warned that upper-class men "too frequently became affected from intercourse with females whose virtue is as loose as their habits are dirty." He added, "Some writers have attempted to prove that the head-louse varies according to the races of men to which it is attached," a point somehow made less often about pubic lice.[8] At the beginning of the twentieth century, another moralist condemned "the modern woman" wearing wigs "cut from the heads of peasants in foreign lands," thus catching their foreign head lice.[9]

In this mindset, dirtiness and filth were considered to breed immorality and criminality. After the articulation of germ theory in the 1870s and 1880s, some concluded that a dirty house was a kind of murder, particularly after bacteriologist Charles Nicolle, of the Pasteur Institute in Tunis, proved that the vector for epidemic typhus was the human body louse. In 1903, Nicolle observed that lice-ridden patients suffering from typhus infected only those working in admissions, not the nurses in the wards. The driver of the disease, therefore, had to be the lice in the victims' clothing, destroyed when the clothing was washed during the patients' hospital stay. After some experimentation, Nicolle published his findings in 1909, and he won the Nobel Prize in 1928.[10]

Typhus seems a relatively new epidemic disease, although some authorities trace its ravages back to ancient times. It may have been endemic in some European populations, but it was not recognized as a separate disease until the sixteenth century and is still hard to diagnose because of its similarities to other diseases. Early modern commentators reported typhus's first appearance during the Spanish siege of Moorish Granada between 1489 and 1492, and it seems to have become more virulent in its transmission to Native Americans and its return to Europe with the conquistadores. After 1500, typhus increasingly accompanied war and most famously decimated Napoleon's Grand Army in its invasion of Russia, reducing it from half a million to only a few thousand.[11]

Typhus is a profoundly disgusting disease. The modern historian Sir Richard Evans describes its symptoms: "violent heats in the body, . . . an unnatural and fetid breath, . . . ineffectual retching, . . . small pustules and ulcers, . . . [and] agonies of unquenchable thirst."[12] The disease was named in the nineteenth century from the ancient Greek *typhos*, meaning smoky or hazy, the state of mind of the typhus patient in the last stages. Until the mid-1800s, it was often confused with typhoid fever, which has similar symptoms but is caused by contaminated water or food rather than the typhus bacterium. Typhus killed 10 to 60 percent of its victims, most of them desperately poor or living in intolerable conditions caused by war, poverty, or persecution. It is often a disease of cold weather, thriving in the layers worn to keep warm, found in modern America only among the bundled-up homeless.[13]

Death by typhus is repulsive; in a way, it is akin to history's imaginary lousy disease, with its suppurating sores hosting hordes of lice. Accordingly, in the nineteenth and twentieth centuries, the insult "lousy" became short-hand for disgusting. The term gained wide use during World War I, notes *The Dictionary of Slang and Unconventional English*. Combatants expected a brief heroic adventure, punctuated by idyllic picnics on the Somme. Instead, they found lice, rats, body parts, and death. As the twentieth century increasingly valued antiseptic cleanliness, its wars were morasses of lice and rats.[14]

After World War I, lice and typhus were targeted by politicians and scientists, including Hans Zinsser, an American physician who had observed typhus's devastation during the war and later wrote *Rats, Lice and History*, about diseases carried by vermin. (Typically, Zinsser was not without prejudices, explaining that he could obtain a sample of lice only by having the police arrest "a colored gentleman who was the only individual easily discovered who was in possession of the coveted insects."[15]) Zinsser warned that while the louse might seem confined to the poor and distressed, or to "primitive" regions of the world, it would "never be completely exterminated, and there will always be occasions when it will spread widely to large sections of even the most sanitated populations."[16]

So far, he's been right. Body lice plagued armies through World War II, and head lice still infest schoolchildren. In the twentieth century, governments tried to loosen lice's hold on their citizens' hair and bodies by limiting immigration and supporting new insecticides. Businesses encouraged the use of DDT until it was banned in the United States in 1972 and the United Kingdom in 1984. In recent times, other treatments have addressed the challenge, and lice have been commodified by businesses small and large, but their appearance still causes children to be ostracized and media to warn of infestations in homeless shelters.

Lice remain a marker of social disgrace and otherness. And, too easily, disgust at the insect extends to the people carrying it. The twentieth century saw active campaigns against both.

Lice and Racism

The debate over race and lice that began in the eighteenth century persisted in the centuries following. To the question of whether different races had different kinds of lice, and thus whether all humans belonged to the same species, answers were often racist. Science and even simple empirical observation were often inflected by cultural assumptions. In the eighteenth century, William Cauty was ambivalent on the relation of lice to race, but he was an exterminator, not a scientist, whatever he might have claimed. Colonists in the Americas were not so judicious. Edward Long, of the white elite of Jamaica, argued in 1771 in *The History of Jamaica* that African slaves had their own black kind of lice, and "this particular circumstance I do not remember to have been noticed by any naturalist."[17] By 1799, the naturalist Charles White cited Long on lice, maintaining, "Perhaps this apparently trivial circumstance may be deemed no inconsiderable arguments in support of the opinion that Africans are a different race than Europeans."[18]

In *The Descent of Man*, written in 1871, Darwin echoed the idea, musing, "The fact that the races of man being infested by parasites, which appear to be specifically distinct, might fairly be urged as an argument that the races themselves ought to be classified as distinct species."[19] In 1845, Darwin had asked the geologist Charles Lyell to collect specimens of lice from enslaved "Negro" persons to show whether their lice were indeed larger and blacker than the European variety.[20] Sending samples of these lice to the entomologist Henry Denny in 1865, Darwin asked, "Will you excuse me asking you to inform me whether *Chiloe pediculi* form a distinct species or a well-marked variety?"[21] Apparently, this question had been on his mind since he sailed on the *Beagle* in 1834. Darwin was familiar with this idea from the work of the entomologist Andrew Murray, who had collected specimens of lice from Africa, Australia, and South America that appeared to differ in both color and structure. He believed that "With insects slight structural differences, if constant, are generally esteemed of specific value; and the fact of the races of man being infested by parasites, which appear to be specifically distinct, might fairly be urged as an argument that the races themselves ought to be classified as distinct species."

Denny's reply to Darwin illustrated how questions of status were infested with lice. He began by insisting, "I cannot see any reason why the Lice of one Human Being should not live on another." The explanation for the natives' relative lack of lice is that they "wear no clothes whatsoever and the women shaved their heads. . . . The men although they wear long hair, so besmeared it with fat and red ochre, as to render it unfit for the abode of even a louse."[23] In this view, dirty hair, unlike the plica polonica discussed in chapter 4, prevents lice instead of producing them. But the two views clearly share the same attitude: foreigners and lice have an intimate—and highly repulsive—connection.

Writing about lice, Henry Denny did not confine himself to South American natives. In his 1842 opus on parasitic insects, he echoed the views of earlier colonial observers about the lice-eating propensities of Native Americans and Africans, proclaiming,

> These creatures however are not regarded as unwelcome visitors by all
> nations, since we are told that the Hottentots and other nations of Western
> Africa, as well as some of the American Tribes eat them, and are so well
> pleased with their dainty morsel that they not only collect them themselves,
> but employ their wives in the chase, and have thence been called Phthirophagi;
> Dr. Richardson informs me that during the overland expedition under Sir.
> J. Franklin, he "daily observed the Indian women cracking their parasites
> between their teeth with much apparent enjoyment." Monkeys have the same
> propensity.[24]

Africans and American Indians, it is clear from this passage, are like mon-
keys. Moreover, the appetites of these exotic and disgusting others are linked
with gender and pleasure. Like the beggars, who the sixteenth-century humanist Heinsius explained take a "fricative pleasure" in scratching their
itches, the natives enjoy their wives feeding them lice.

Similar reports of the taste for lice appear throughout the later nineteenth
century. The American entomologist A. S. Packard argued that "the Chinese
and other semi-civilized people" eat lice and that consumption leads to
"certain mental traits and fleshly appetites."[25] Mark Twain, who had his own
racism issues, wrote his mother in 1862 that the Indian "Hoop-dedoodle-do"
unwillingly sheds prodigious amounts of vermin that otherwise he would
eat, knowing "something about them which you don't; viz, that they are good
to eat."[26] The question of consumption continued to absorb observers in
the early twentieth century, when the specter of vermin was used to justify
imperialism, nationalism, and anti-Semitism. The German entomologist
Heinrich Fahrenholz (1882–1945) argued that not only did Black people have
their own species of lice, but so did the Chinese, the Japanese, and the Native Americans.[27]

Lice are bad, but people are worse. The story of lice in the twentieth century is a daunting record of human depravity. At the same time that science
discovered that lice spread typhus and other diseases, the mythology of the
lousy was equally lethal to millions. Lice-borne typhus killed hundreds of
thousands, particularly in Eastern Europe, but the insect's role as a disease
vector was at least natural. The use of lice to justify the extermination of those
considered parasitic was a more unimaginable horror. Reichsfürer Heinrich
Himmler, the architect of the Nazi death camps, proclaimed in 1943, "Anti-
Semitism is exactly the same as delousing. Getting rid of lice is not a question
of ideology. It is a matter of cleanliness. In just the same way, anti-Semitism,

"The Jews are lice: they have typhus," reads this Nazi propaganda poster from 1942. From the collection of the District Museum in Rzeszów, https://encyclopedia.ushmm .org/content/en/photo/antisemitic-poster-published-in-german-occupied-poland-in -march-1941

for us, has not been a question of ideology, but a matter of cleanliness, which now will soon have been dealt with. We shall soon be deloused. We have only 20,000 lice left, and then the matter is finished within the whole of Germany."[28] In a Nazi 1942 propaganda poster, the image of the Jew and an image of the insect are intermingled, breaking the border between louse and human, both assaulting the safety of the threatened human viewing the picture. Such a threat justifies the elimination of the dangerous inferior creature by smashing or gassing.

Hugh Raffles, a historian of the Holocaust, argues that Himmler probably thought Jews and lice were literally the same.[29] But whether the Nazi leader's words were meant to be metaphoric or literal, he was manipulating a centuries-long association of Jews with disease, going back to accusations of well poisoning during the fourteenth-century Black Death and the descriptions of the plica polonica in the early modern period. Himmler was also invoking the recent demonstration of lice's link to typhus, putting it to a use never imagined by Charles Nicolle.

In earlier times, insecticides were made from ingredients ranging from pig's grease to mercury. By the early twentieth century, it was common to use Zyklon B, a chemical in the cyanide family, against the bugs. American officials used it to delouse immigrants, particularly those coming from Eastern Europe and the Far East, as well as Mexican day laborers crossing into Texas. Using it in the death camps, Himmler expressed gratitude to the Americans for testing its efficacy.[30]

During and after World War II, DDT often performed lice-killing duties, and it was widely sprayed on soldiers and refugees—too late for Anne Frank and her sister, who died of typhus in February or March 1945 at the Bergen-Belsen concentration camp. A surviving prisoner recounted of Frank that at her death "She was no more than a skeleton by then. . . . She was wrapped in a blanket; she couldn't bear to wear her clothes anymore because they were crawling with lice."[31]

Today, Western Europeans and Americans are no longer threatened by typhus, which can readily be cured by a shot of antibiotics, although it continues to kill many thousands in the developing world. (There were fifty thousand cases in Burundi during its 1980s civil war.[32]) But the fear and disgust generated by lice are still with us. Despite its reduced impact as an entomological grim reaper, the insect itself still causes a reactive horror. Head lice, which actually do not carry any kind of illness (although some nonmedical lice hunters dispute this), are considered a kind of plague endangering civil society, not to mention the tender heads of our children.

Head lice can turn children into pariahs. The unfortunate families so infested feel overwhelming embarrassment and humiliation. In a recent episode of the situation comedy *Modern Family*, two characters discuss the news that lice has been found in their Vietnamese daughter's class. "Ugh," says one, "it's probably from Portia. You know, she's always so filthy. They had to kick her out of Swim Buddies because she left a ring around the pool." When lice are actually found on their daughter, they decide not to tell anyone—no one will want to have anything to do with a child, or her family, that has lice.[33] The parents involved happen to be gay. In the twenty-first century, having lice is apparently more awkward than homosexuality or race. Social attitudes have evolved, but not enough to tolerate lice.

Lice Go to War

The British soldier/poet Isaac Rosenberg, soon to be killed at the Battle of Arras in 1918, wrote the poem "Louse Hunting":

Nudes—stark and glistening,
Yelling in lurid glee. Grinning faces
And raging limbs
Whirl over the floor on fire.
For a shirt verminously busy
Yon soldier tore from his throat, with oaths
Godhead might shrink at, but not the lice.
And soon the shirt was aflare
Over the candle he'd lit while we lay.
Then we all sprang up and stript
To hunt the verminous brood.
Soon like a demons' pantomime
The place was raging.
See the silhouettes agape,
See the glibbering shadows
Mixed with the battled arms on the wall.
See gargantuan hooked fingers
Pluck in supreme flesh
To smutch supreme littleness.
See the merry limbs in hot Highland fling
Because some wizard vermin
Charmed from the quiet this revel
When our ears were half lulled
By the dark music
Blown from Sleep's trumpet.[34]

A constant refrain in firsthand accounts of World War I is the omnipresence of lice, which torture the soldiers and, as Rosenberg's poem recounts, turn them into gibbering shadows writhing in a kind of demonic ballet. Some of the themes in this poem are familiar from earlier warfare. During the American Civil War, soldiers were tormented by lice and fleas. But Rosenberg's account of lice hunting during the Great War is more visceral—the soldiers did not merely strip off their clothes, but they became monsters themselves, with "gargantuan hooked fingers" that sought to "smutch supreme littleness." Lice hunting became a cosmic metaphor for the battle between man and beast, with the lice triumphant as man shook off any spark of Godhood in his pursuit of the enemy burrowing into his skin.

Not all soldiers had such vivid memories of nightly encounters with the "wizard vermin," but no account of the war omitted the tormenting inescapability of lice. The trenches teemed with lice (and rats), and soldiers and the nurses who cared for them rarely had the chance to change their clothes or bathe. Another poem, by Siegfried Sassoon, catches the profound despair induced by lice incursions: "In winter trenches, cowed and glum, / With crumbs and lice and lack of rum, / He put a bullet through his brain. / No one spoke of him again."[35]

For a generation of young men, and some women, brought up in the nineteenth century's new era of cleanliness, the degradation of war was deeply destabilizing. It could lead to suicide, or as Evadne Price (writing as Helen Zenna Smith) wrote about a female ambulance driver in the novel *Not So Quiet . . . Stepdaughters of War*, "Her soul died under a radiant summer moon in the spring of 1918 on the side of a blood-splattered trench."[36] Earlier in the book, the ambulance drivers watched as one of their number cut off her hair because of the "squadron of lice" that had taken residence on her head: "Snip, snip, snip," and her tresses fall. Another comrade warned, "You'll look awfully unsexed." But the loss of femininity, and of gender roles, are worth it in the pursuit of lice, a symbol for the conflict's degradation and loss of civilization.[37]

Price's book was supposed to parody the most famous World War I novel, *All Quiet on the Western Front* by Erich Maria Remarque, although it is a sample of grotesquery and black humor rather than laugh-out-loud funny. *All Quiet* is equally steeped in the despair produced by war, and by lice. The narrator, Paul Bäumer, recounts, "Killing each separate louse is a tedious business when a man has hundreds. The little beasts are hard and the everlasting cracking with one's fingernails very soon becomes wearisome." The soldiers sit around naked, throwing their prey into a tin on the fire where they crack and die.[38] Even to a non-literature major, it's clear that the lice's fate also awaits the young soldiers.

Not surprisingly, soldiers would try anything to get rid of these demeaning pests, from the lice roasts just described to anti-lice products eagerly offered by entrepreneurs. Perhaps the most ingenious was the body cord, an

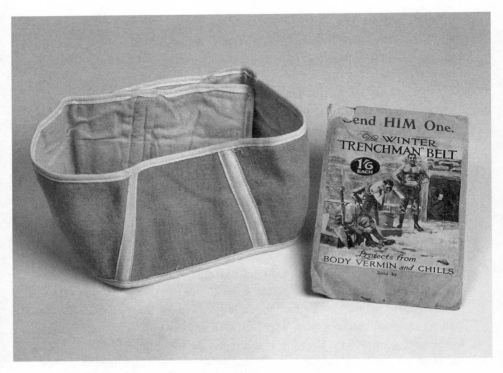

Trenchman belt. Courtesy of the Imperial War Museum

insecticide-laden belt that infested soldiers—or ambulance drivers—wore around their middle. According to the manufacturer, "The skin absorbs its germicide properties, and these are carried to all parts of the body. It even prevents the lice from lodging in the clothing."[39]

The Asiatic Body Cord was endorsed by a lady who presumably knew how to keep a young man clean, sweet, and virtuous. In its advertisement she exults, "Have just received word from the trenches that Somerville's Body Cord is the finest thing ever invented for a soldier's comfort."[40] The cord was inspired by a device from Indian folk medicine; it was made of three strands of four-ply robe, plaited together and doused with a mixture of two parts mercury ointment and one part yellow beeswax. During the war, 120,000 belts were sold, although the medical establishment doubted their value.[41] This product seems to have benefitted from the same assumptions held by the customers of John Southall's bedbug elixir: an exotic ancient culture—especially one deep in bugs—might know more about destroying insects than Western medicine. Likewise, as the conditions of the battlefield threw the combatants into a uncivilized state, a product inspired by a supposedly uncivilized culture might work. Another marketed lice killer was the Trenchman Belt, shown on the facing page.

On the package, two soldiers scratch at their lice as a virile young man stands upright and grins while wearing his Trenchman Belt, which will protect him from not only lice but also colds. The belt was a triumph of British advertising, if not British medicine. More soldiers, however, depended on Keating's Powder, made from pyrethrum, a chemical derived from chrysanthemum leaves, sold in a can that featured a grinning little devil, reflecting the bugs' satanic qualities. Its effectiveness is unknown, but pyrethrum is still the active ingredient in the modern lice shampoos to which lice have become resistant—possibly because of overuse. Keating's Powder was also called Persian Powder, another nod to exoticism to sell something to desperate Westerners.

The only real weapon against lice was cleanliness—essentially impossible in the trenches. When not deployed, the soldiers sometimes had showers, although the hot water necessary to kill the vermin was also scarce. Their uniforms, including underwear, needed a steam wash, but even if the afflicted was lucky enough to have himself and his kit cleansed of lice, he would almost immediately be reinfected at the front. Some soldiers declared that they would prefer enemy fire to the intense itch of lice bites; according to one military historian in 1915, "The irritation due to the body-louse weakens the host and prevents sleep, besides which there is a certain psychic response which causes many officers to fear lice more than they fear bullets."[42]

Lice spawned many synonyms—Americans and the British called the creatures "cooties," enlivening the insult range of generations of schoolchildren. Americans also referred to them as "graybacks" (a term from the Civil War) or, with considerable poetic license, "galloping dandruff" or "seam squirrels." The British also named them "coodlers," while the French, no doubt with twirling mustaches, called them "totos." The English and Australians also referred to them as "chatts," perhaps from the chatting that accompanied nitpicking. The nighttime terrorists, many soldiers also told each other, were "made in Germany."[43]

Although lice were constant on the Western Front, typhus was less common. But another lice product, the less lethal trench fever or Rickettsia quintana (sometimes called relapsing fever), was rampant. Relapses could occur numerous times after the first bout. Like typhus, it produced high fevers, sore muscles, and skin lesions, but it was also usually accompanied by sharp shin pains. It could lay up a soldier for three months, possibly preventing death in the trenches but also potentially leading to a lifetime of depression. Trench fever affected J. R. R. Tolkien, A. A. Milne, and C. S. Lewis, who all returned home to create fantasy places—the dark world of Mordor, with its resonances of the trenches; the pastoral landscape of the Hundred-Acre-Wood; and the religious utopia of Narnia. Perhaps Shelob the Spider in *The Lord of the Rings* entomologically evokes the lice of the trenches, as may A. A. Milne's reaction to an insect zoo in London: "It makes me almost physically sick to think of that nightmare of mental and moral degradation, the war. When my boy was six years old he took me into the Insect House at the Zoo, and at the sight of some of those monstrous inmates I had to leave his hand and hurry back to the fresh air."[44]

As much as the soldiers on the Western Front suffered from lice and associated diseases, things were far worse on the Eastern Front. Typhus rolled east from Serbia in 1914, following the Austrian invasion. (The Serbians blamed the disease on "Albanian lice.") It sickened almost 500,000 Serbian troops along the front, killing 120,000. The American doctor and bacteriologist Hans Zinsser went to Serbia in 1915 as a member of the Red Cross Typhus Commission, observing the carnage firsthand. He returned in 1917–19 as a member of the Medical Corps of the US Army. Zinsser's bacteriological labors in the 1930s, working with fellow scientist M. Ruiz Castanada, led to the discovery of antibodies in the blood serum of typhus patients. Ultimately, by inoculating normal chick tissue with Rickettsia grown in chick embryo yolk sacs, they produced a vaccine containing dead bacteria, initiating an immune system response that curtails the disease. Zinsser also identified Brill disease, a recurrent, less virulent strain of typhus that was renamed Brill-Zinsser disease in his honor.[45]

Based on these experiences, in 1935, Zinsser published *Rats, Lice and History*, dedicated to his close friend Charles Nicolle, calling it a "biography" of typhus. As one of the first historians of science (although he modestly refrained from calling himself that), Zinsser recounted the role of parasitic diseases in human history, something professional historians had failed to recognize. More importantly, defining his work as a biography opened up a new way of envisioning disease and its meaning. Lice became a metaphor for many other things—including, in Zinsser's interpretation, those who give up freedom for the security of an easily accessible food source. For lice, the nutrient is human blood. The louse achieves "a secure and effortless existence on a living island of plenty. In a manner, therefore, by adapting itself to parasitism, the louse has attained the ideal of bourgeois civilization, though its methods are more direct than those of business or banking, and its source of nourishment is not its own species."[46]

Zinsser, a humane man, would have been horrified to see the Nazis later employ a similar metaphor of the bloodsucking louse to justify genocide. But his example shows the explosive possibilities of verminous associations. Lice are parasites, Zinsser argued, but so are men, who depend for their existence on exploiting the rest of nature. And pity the louse, he says (presumably facetiously), who like us is prey to typhus and dies from the disease as we do: "If only for his fellowship with us in suffering," Zinsser said of the creature he studied, "he should command a degree of sympathetic consideration."[47] As with Heinsius in the seventeenth century, lice became a way of commenting on the human condition.

As part of a League of Nations commission investigating a typhus epidemic, Zinsser visited the Soviet Union in 1923 and was as little impressed with their system as with the rabid capitalism he deplored. "The governing mob," he wrote, "cared little in those days about a hundred thousand lives more or less, starving children, suffering and sickness, if only they could attain the noble ideals of Marxian theory."[48] In fact, Lenin had understood the louse's threat to the Russian Revolution, declaring, "Either socialism will defeat the louse, or the louse will defeat socialism."[49]

By 1923, Russia had suffered approximately twenty-five to thirty million cases of typhus after the 1917 Civil War between the Reds and the anti-Communist Whites. Three million people, mostly civilians, died from the disease. Like so many others, the Soviets linked the louse to the despised other, in this case their White opponents. A poster issued in 1921 proclaimed, "The Red Army has crushed the White Guard parasites—Yudenich, Denikin, and Kolchak [White military leaders]. Comrades! Fight now against infection! Annihilate the Typhus-bearing louse!"[50] In the illustration, women cleanse a lice-ridden man, killing the lice in his clothes and then washing them under the direction of a military doctor. Unfortunately, none of this—if it ever happened—achieved much progress against typhus.

Western Europeans and Americans closely observed the devastating effects of typhus in Eastern Europe. During the war, Germany issued a postcard emphasizing the barbarism of the Russians. The text explains that family members are delousing themselves, a practice the German soldier in their midst is forced to follow. The Russians live in filth with their animals, including a pig rooting on the floor. In his Gresham Lecture "The Great Unwashed,"

Sir Richard Evans argued that the German government used the image to show the "fundamentally backward and uncivilized" nature of Russian culture.[51]

Germany and Austria delayed invading Serbia until the typhus epidemic there had run its course, and Germany took measures to prevent an outbreak among its citizens and soldiers, especially those sent to the Western Front. Anyone showing symptoms was put into quarantine, and huge delousing plants were set up on the border. Strict standards of hygiene were instituted. These efforts were successful; of the thirty-three thousand German Army deaths from infectious diseases, only fifteen hundred were from typhus.[52]

Nevertheless, the psychological effects of the typhus epidemic, especially the Soviet experience during the Civil War, were shattering. Some Westerners refused to believe a German Red Cross relief team's reports that typhus and war were driving people to cannibalism and that dogs were eating corpses. The deaths included a large percentage of the old-guard aristocrats and the intelligentsia.[53]

Winston Churchill believed that the Germans had sent Lenin to St. Petersburg in 1917 to destabilize the enemy, "the way you would send a vial containing a culture of typhoid or cholera to be poured into the water supply of a great city." Indeed, according to Churchill, Russia was a country of "armed hordes smiting not only with the bayonet and with cannon, but accompanied and preceded by swarms of typhus-bearing vermin which slay the bodies of men, and political doctrines which destroy the health and even the soul of nations."[54]

Churchill knew his lice. As a colonel at the front in 1916, he greeted his officers with the words "War is declared, gentlemen, on the lice." One of his officers, Andrew Dewer Gibb, later related, "With these words was inaugurated such a discourse on the *pulex Europaeus*, its origin, growth and nature, its habitat, and its importance in wars ancient and modern, as left one agape with wonder at the force of its author."[55] If Napoleon could have anticipated these remarks and the disease that lice carry, perhaps Russia would have fallen into French hands in the early nineteenth century and the Great War avoided altogether. As a man of action, not just words, Churchill had brewery vats converted into bathtubs and initiated a general delousing. It worked; he gained his men's respect, and they escaped the depredations of typhus and trench fever.[56]

American authorities also took a hard line on lice, closely examining Eastern European immigrants at the turn of the century. Even before boarding the ships to New York, they were housed in "concentration camps" in Britain that were maintained by the shipping companies, where they were inspected and treated for lice. If found to be lousy, men had to shave their hair; women, to preserve their "chief glory," were allowed to use soap, oil, and lice combs against the insects. On arrival in New York, they were inspected again. From Ellis Island, steerage, third-class, and some second-class passengers found to have lice were sent to a separate delousing station on Hoffman Island, where they were treated with oil and soap and their baggage blasted with either pressurized steam or cyanide gas.[57]

American officials insisted the process was conducted with respect. The *New York Times* assured its readers, "It is felt that the method of handling those who undergo this process is calculated to maintain, as much as is compatible with thoroughness and efficiency, the dignity and pride of the individual."[58] But the linking of immigrants with disease helped drive anti-immigration fervor. In 1892, a typhus outbreak on the Lower East Side of New York was traced to Jewish immigrants who had recently arrived on the steamship *Massilia*. Although only two hundred people were infected, the authorities temporarily detained all Jewish immigrants and sent health inspectors into the immigrants' boarding houses. The action quickly progressed to unnecessary brutality. According to the *New York World*, health inspectors "carried away women while their husbands tore their hair and children wept in frightened ignorance. It was a dreadful task, for all the patients were ignorant and already cowed by oppression. They were being hurried away to execution for all they knew."[59]

William Randolph Hearst capitalized on the fear with a terrifying picture of a louse above a panic-inducing headline: "NEW DISCOVERIES ABOUT 'COOTIES,' THE SOLDIER'S PEST," it warns, adding in subheads, "Why Scientists Are Studying So Patiently Every Obtainable Fact about Them"; "All Our Returning Troops Are So Carefully Deloused"; and "How the Arrival of a Few War Cooties Here Might Sweep America with a Death Plague That Kills Four Out of Every Five of Its Victims." In this instance, returning soldiers (particularly Black soldiers) are labeled possible lice carriers, but immigrants also could be tarred with the same brush (or tentacles).[60]

Even a publication as staid as the *Public Health Bulletin*, from the newly created Public Health Service, was not immune to panic, especially after the US surgeon general, during his inspection of Eastern Europe in 1922, "had the pleasure of this intimate personal contact with friend cootie." Abandoning jocularity, the *Bulletin* warned, "Typhus is just as much an international problem at this moment as the situation in Belgium was while the Germans occupied the country. It is a menace to the whole world." By 1922, the US

Public Health Service had taken control over foreign entry into the United States, and its agents were charged with investigating immigrants and ships' crews for "loathsome, contagious, and chronic diseases." (The language comes from the 1891 Immigration Act, enacted in response to a recent cholera epidemic; "loathsome" in this case has both a physical and moral meaning.) Agents operated not only in New York and other eastern cities, but also along the Mexican border and in San Francisco, where two people died from cyanide poisoning aboard the SS *Tahiti*.[61]

The most chilling American example of the identification of foreign lice with foreign people surfaced in 1917 in El Paso, Texas. A typhus outbreak in central Mexico in 1915 caused some officials to panic. The mayor of El Paso requested the US Customs Service to set up a delousing facility to "bathe and disinfect all the dirty, lousy people who are coming into this country from Mexico." The Mexicans were bathed with gasoline, stripped, and inspected before being allowed to proceed to their jobs. One Mexican maid, Carmelita Torres, refused to undergo this humiliating treatment and encouraged thirty others to also refuse this initiation into American society. The ensuing "Bath Riot" was ultimately put down by contingents of the American and Mexican armies.[62]

Afterward, delousing became a permanent part of crossing the border until the late 1950s, reflecting the ingrained bias that Mexicans were dirty—even though, somewhat ironically, most of the Mexican women crossing the border were coming to clean American homes. Zyklon B soon replaced gasoline as the fumigating agent of choice, and it was in turn replaced by DDT in the 1950s. Increasingly, to avoid these demeaning procedures, immigrants from both Mexico and China started entering the country surreptitiously. It might not be an exaggeration to say that lice created illegal immigration.

On both sides of the Atlantic in the 1920s and 1930s, scientists labored to find a vaccine against the ravages connected to lice. Hans Zinsser was one of the scientists laboring on a vaccine for typhus in the 1920s and 1930s. Another was a Polish bacteriologist of Austrian descent, Rudolf Weigl. Like Robert Hooke and Antonie von Leeuwenhoek before him, Weigl used himself as his first subject, both contracting and surviving typhus. Weigl continued his research when the Nazis occupied Lvov (modern Lviv), ultimately developing a vaccine, while his laboratory became a refuge for Polish intellectuals, Jews, and members of the Polish underground. He employed "lice-feeders" to provide nourishment for the millions of lice that he needed—perhaps not a glorious job, but one that provided a way to escape the Wehrmacht, who were terrified by the disease and those who might carry it. Their fear was so intense that a Jewish bacteriologist, Ludwig Fleck, who had worked with Weigl, was allowed to continue his research into a typhus vaccine, first at Auschwitz concentration camp and then at Buchenwald.

The horrifying mix of lice, typhus, and genocide permeated the concentration camps. When prisoners arrived, they were deloused in a particularly dehumanizing procedure—they were forced to strip, thoroughly shaved—including pubic hair—and then doused with chlorine. As one survivor recalled, the process was as humiliating as possible:

> Then when we were undressed, we were ordered, everybody was ordered to stand up on a stool, and they shaved us, they shaved our hair, and the private parts, and we looked, we couldn't even recognize each other once we were stripped, not only of our clothes, but of our hair. Then we were shoved into those, um, showers, and they first opened the hot water, so we were scalded and as we ran out from under the hot water, we were beaten back by the SS and by the Kapos to go under the showers again, so they opened the ice cold water, which had the same effect, and finally we were out of this shower.[63]

Paul Julian Weindling, a historian of the Holocaust, writes, "The ordeal of delousing was as much a psychological as a physical torment." The racialized understanding of typhus, he adds, provided an additional rationale for the killing of the "human vectors" of the disease.[64] The brutality of the guards expressed the linked repugnancies of Jews and typhus, and the fear that the weak can somehow kill the strong, that the lice of the prisoners would somehow escape and kill their persecutors. The guards at Auschwitz warned their victims, *Eine Laus dein Tod* ("One louse your death").[65] Since it was virtually impossible to avoid lice in the camps, it is easy to see how death was always imminent.

Primo Levi, who survived Auschwitz to become a famed novelist of the Holocaust, has argued that "Here was not only death but a host of maniacal and symbolic details, all intended to demonstrate and confirm that Jews, and Gypsies, and Slavs are beasts, fodder, garbage. . . . The very method chosen (after careful experimentation) for extermination was openly symbolic. The same poison gas employed for disinfecting ships' holds and rooms infested by bedbugs or lice was to be used, and was used."[66]

Maniacal indeed was how deeply lice informed and even shaped the Nazi determination to eradicate the Jewish parasite. The gas chambers at Auschwitz were designed to look like delousing facilities, sometimes with fake showerheads and signs that announced *zum baden* ("to the bath") and *zur Disinfektion* ("to disinfection"). Historians argue that the Germans invented this deadly charade to pacify their victims, but the entire history of lice also fueled the elaborate masquerade. Jews were lice requiring extermination. The medical technicians pumping the poison into the gas chambers were called "disinfectors"; the Zyklon B containers bore the warning "Cyclon, to be used against vermin."[67]

Holocaust deniers have argued that there were no Nazi death camps, and that hydrogen cyanide was used for delousing, not genocide. According to Friedrich Paul Berg of the Institute for Historical Review, a Holocaust-denying organization, "The purpose of the delousing chambers was to save lives—and that is not denied except by the most passionate Extermination-ist. No doubt, many hundreds of thousands of people, possibly millions, in-cluding countless Jews, owe their lives to these chambers and the German technology based upon Zyklon-B."[68] Deniers maintain that the war on ty-phus, not the annihilation of the Jewish population of Europe, produced Auschwitz, Dachau, and the other concentration camps. Nazis were not mur-derers, but simply the most efficient lice exterminators.

These claims have been sufficiently debunked. But they show how the fear of lice, and of lice-borne disease, can be used to cover the most extreme de-struction. Even the heroes in the war against lice saw the enemy as not just an insect but as human savagery. Charles Nicolle, in his Nobel Prize accep-tance speech in 1928, fused lice and the people who had them: "Man carries on his skin a parasite, the louse. Civilization rids him of it. Should man re-gress, should he allow himself to resemble a primitive beast, the louse begins to multiply again and treats man as he deserves, as a brute beast."[69] Infesta-tion is not just physical, but moral.

Just before World War II, the civilized world found another tool against lice and typhus. DDT, a chemical synthesized in the late nineteenth century, was identified as an insecticide in 1939 by a Swiss scientist, Paul Hermann Müller. By the end of the war, the Allies were using it in a successful effort to delouse soldiers and civilians. It would soon be employed as a boon for com-mercial enterprises as well as the military.

Postwar Lice Fighting

After the war ended, businesses saw at least one benefit from the carnage: the general use of DDT as an insecticide. Horrifying to modern sensibilities, ad-vertisements touted the chemical's benefits. One ad for DDT suggested that

Advertisement for Trimz DDT wallpaper, *Women's Day*, June 1947. Courtesy of the Science History Institute, https://digital.sciencehistory.org/works/mg74qm295

being a good mother required DDT-impregnated wallpaper. Because insects breed in "filthy" homes, morally upright housewives needed this product. Using DDT would also mean a happy baby. Finally, an omnipotent authority endorsed DDT: "Tested and commended by *Parents' Magazine*."

In 1962, Rachel Carson's *Silent Spring* showed that DDT was not benign as the chemical companies promised, and its dangers to the environment were substantial. She also showed that in many parts of the world, lice were rapidly developing resistance to DDT. The assistant director of the Agricultural Research Division of the American Cyanamid Company responded that Carson was "a fanatic defender of the cult of the balance of nature" and that if her advice were followed, "we would return to the Dark Ages, and the insects and diseases would once again inherit the earth."[70] This critique has been echoed ever since, as when the conservative scholar and political commentator Steven Hayward declaimed in 2014, "Few books since *Das Kapital* have done more damage to humans than *Silent Spring*, and yet she—and her dreadful book—continue to be honored by the Left."[71] While the current debate about DDT focuses on its role, circumscribed or not, in combatting mosquitoes and malaria, the passion shown by both proponents and adversaries of the pesticide demonstrates its powerful social and political resonances.

The emotions raised in the battle against lice is most evident in the controversy over whether children found to have nits or lice should be allowed to stay in school. Until recently, schools in the United States and the United Kingdom had a "no nits" policy, ejecting children from the classroom as soon as the pests were discovered. According to most medical and public health authorities, head lice, unlike their bodily cousins, carry no diseases. Nevertheless, head lice have absorbed many of the stigmas of body lice, including the association with foreigners and filth. During the recent influx of immigrant children from Mexico and Latin America, one health care worker reported, "I would be talking to the children and lice would just be climbing down their hair."[72] Then she claimed that the authorities had forbidden her to talk about it. Although the children traveled on chartered buses and planes, Todd Starnes of Fox News warned, "I don't mean to upset anyone's Independence Day plans, but were these kids transported to the camps before or after they were deloused? Anyone who flies the friendly skies could be facing a public health concern."[73]

"A public health concern" about head lice infestation (pediculosis, in medical terminology) is hardly the same as the threat from Ebola or HIV or COVID-19, but the reactions of some parents often border on the hysterical. Sometimes, the school nurse is blamed. "Repeatedly," reported a letter to the editor of the *American Academy of Pediatric News*, "parents become volatile to the extreme of making death threats to school nurses and school staff over this issue."[74] An exchange on the *New York Times* opinion page spurred many comments.[75] When the Harvard entomologist Richard Pollack pointed out that neither the American Academy of Pediatrics nor the National Association of School Nurses supports removing children with head lice from school, he was accused of having a conflict of interest and of having no personal knowledge of lice. (Because, like many lice professionals before him, Pollack often acts as a lice feeder, this charge is easily—and itchily—dismissed.[76]) Others argue that lice can come only from dirty homes and dirty children, and, echoing ancient theological arguments, that lice serve to make filthy people clean their homes. But lice do not discriminate between clean and dirty hair, and if they do, they probably prefer the cleaner variety. Another parent, mystified over the passion expended on the lice issue—a nonissue in Asia where she grew up—finally realizes "the reaction to lice was a cultural thing!"[77]

It is also a class thing. Some people argue that no-nits policies cost the children learning time in school, especially if they are forced to stay home until there is no evidence of lice or nits, and cost their parents time at work and therefore income. The removal of the bugs is also enormously time consuming, involving a close combing of hair—particularly burdensome when active lice scurry out of the way or because nits cling with special tenacity to the hair shaft. When that happens, lice wars evolve into a class struggle between those who can afford to pay professional nitpickers—yes, that is a profession—and those who cannot afford the expenses involved in eradicating lice.

And so, as usual, issues of culture and class devolve into issues of money—among their other characteristics, lice are lucrative. The amount of money involved is so large that it has engendered a battle among Big Pharma, environmentalists, and advocates of natural remedies. At its center is lice's increasing resistance to pyrethrin, the active ingredient in the lice-killing insecticide Rid, and perhaps to the closely related chemical permethrin, used in Nix, and to other products promising to expel lice from children's hair. Some products, like Malathion and the neurotoxin lindane, have serious potential side effects, including seizures and even cancer. More natural approaches usually involve smothering the lice with some kind of oil or mayonnaise. One American business, catchily named Head Lice to Dead Lice, has produced an award-winning animated video, available for only $39.95, starring a cartoon character named Jana McNanna, who is ostracized at school for having "cooties" and allowed back into school and community only after her mother follows the method for applying olive oil. It even provides a game, similar to Chutes and Ladders, that shows how to apply the oil.

Another expensive tool is the LiceBuster, renamed the AirAllé, which shoots a stream of hot air at the insects and dehydrates them to death. Larada Sciences, which owns Lice Centers of America, has franchised the device to various lice parlors around the country, including Rapunzel's Lice Boutique in Michigan, a name proving that in America the right packaging can make almost anything cool.[78] Other outfits, including the National Pediculosis Society, sell a variety of lice combs. Even scientists have entered the lousy marketplace—one group of professors from Purdue has applied for money to sequence the genome of the head louse, which apparently has the smallest genome of all the hemimetabolous insects.[79] A group of British scientists has established Insect Research and Development Ltd. to evaluate lice products.[80] An American company, Identify Us, provides information about lice and other insects, including identifying submitted specimens and offering pro bono services for those who cannot afford lice defenses.[81]

One species of lice may be giving up the ghost without expensive or dangerous countermeasures. According to Ian Burgess of Insect Research and Development, "Pubic grooming has led to a severe depletion of crab louse populations . . . an environmental disaster in the making for this species."[82] Environmentalists committed to species survival may be pleased to learn that this claim is widely dismissed, because not many of the world's humans indulge in Brazilian waxes, the culprit identified as destroying the habitat of pubic lice—at least among affluent twenty- and thirty-somethings. The charge seems to echo the cultural bias of a previous claim, made forty years ago by the eminent entomologist and historian J. S. Busvine, that an increase in head lice and scabies was a result of hirsute hippies' poor grooming habits.[83]

From the beginning of the twentieth century until today, the common theme of all reactions to lice is disgust. The creatures are invariably associated with the other, whether German soldiers, Russian counter-revolutionaries, Jewish inmates, or dirty next-door neighbors. Lousiness is a state of being that easily glides from the metaphoric to the physical, with responses moving quickly from angry words to furious actions. The sacrificial burning of lice in World War I trenches reflects the annihilation of the veneer of Western civilization during the war to end all wars. In that war, young men, whether pierced by bullets or bugs, are transformed into little more than blobs of bleeding flesh. This transformation was anticipated by Franz Kafka, whose *The Metamorphosis* envisions a protagonist awakening one day to find himself turned into a giant *Ungeziefer*, a kind of crawling, nasty bug or vermin.[84] Nazis used the long metaphoric history of lousiness to justify genocide when they fused the Jewish *Untermensch,* or subhuman, with the louse—both considered spreaders of typhus. They learned how to use Zyklon B from the American example in treating immigrants, although happily in the United States, disdain rarely became lethal. Since World War II, having lice is no longer a death sentence from disease or persecution, but the condition continues to spur horror, far out of proportion to its real dangers. We no longer spray children's rooms with DDT, but many parents with lice-beset children will dare do almost anything to get rid of them, as lice suck up not only blood but also money.

"Don't panic!" urge lice-fighting organizations and school notices, but lice feed on fear as much as blood. Lice get under our skins more deeply than any other insect parasite. The bedbug is currently socially alarming, but it is a driver of discomfort rather than disease. The flea is now a threat more associated with our pets, covered by our general indulgence toward them.

Even lice, although a painful source of disgust and humiliation, are no longer seen as a danger to our entire civilization. Like the fleas we will discuss in chapters 6 and 7, they can be viewed as both frightening and funny.

CHAPTER SIX

The Flea in Humanity's Ear

"Sir," once remarked Samuel Johnson, the eighteenth-century ruler of English literature, "there is no settling the point of precedency between a louse and a flea."[1]

It's indeed hard to choose between the two vermin—for one thing, they're both very small—but there is a lot to say for the flea, in general terms a far more benign parasite. Before medicine revealed that fleas carried bubonic plague, they made people laugh; they were the comic relief of the insect world. Unlike lice, nobody ever thought that God used fleas to punish the mighty, or to mark beggars and foreigners for ostracism. Rather than a satirical snicker, fleas set off belly laughs. They seemed far too inconsequential and ridiculous to ever take seriously. Every group associated with fleas—experimental scientists, the demonically possessed, women, slaves—could be dismissed as laughable. But as with much comedy, the joke was often fueled by contempt—and the uneasy awareness that the next laugh might be on the audience.

Sometimes the derision aimed at fleas was used to justify violence against those identified with the pests. Like the louse, the flea had misogynistic and racist associations. To torture or even kill some human beings seemed "no more than squashing a flea." But the subtext of the assault was not a triumph over the weak, but rather how potentially threatening the weak could be—and thus how necessary it was to control them. This fear particularly resounded in the need to control women and slaves. When Queen Christina of Sweden (1626–89; reigned 1634–54), a ruler known for cross-dressing, shot fleas out of her four-inch cannon, she was making a statement of what she could do to the insignificant masculinity of her male advisors who were trying to tell her what to do. Another royal princess was bitten by a flea while she watched a flea circus, demonstrating the fleas' proclivity to escape captivity and bite their captors—this little pest could be dangerous.

The flea is an extraordinary insect, one of the smallest that prey on animals and humans. There are sixteen hundred species of fleas, including the oriental rat flea (*Xenopsylla cheopis*), the major disease vector for bubonic plague and murine typhus, and the human flea (*Pulex irritans*), now increasingly rare owing to insecticides and modern hygiene. But before knowing the lethal character of *X. cheopis*, humans were captivated by both its tiny size, one-sixteenth to one-eighth of an inch, and its impressive jumping ability. Scientists disagree on how far and high a flea can jump. May Berenbaum, a professor of entomology at the University of Illinois, estimates that a flea can vault a hundred times its own length—in layman's terms, that's equivalent to a human doing a standing broad jump of a quarter of a mile. She dismisses the argument that a flea's jump is equivalent to a man jumping over St. Paul's Cathedral (other biologists use the Empire State Building in their estimates).[2] Still, whatever its ability to leap, the flea's tininess made it a symbol of vulnerability and insignificance.

Aristotle taught that fleas, like lice and other insects, were spontaneously generated from inanimate matter. The similarity of the Latin terms for dust, *pulvis*, and flea, *pulex*, underlined the insect's identity as insignificant. The naturalist Thomas Muffet, who had so much to say about bedbugs and lice, argued that fleas were generated from dirt and sweat, but especially from human and goat urine, and attack most commonly when men are sleeping. Fleas, he judged, are "not the least plague," even though they can be eradicated by numerous herbal remedies and warded off "by a glowworm set in the middle of the house."[3]

Like the louse, the flea was often invoked to comment on human pretension, but with a subtle difference. Lousy people were often considered degenerate and disgusting, and the bedbug infested were viewed as filthy, but the flea bitten (literally and figuratively speaking) were more often considered simply absurd. Muffet explained, "Though they trouble us much, yet they neither stink as Wall-lice [bedbugs] doe, nor is it any disgrace to a man to be troubled with them, as it is to be lowsie." In short, they seem far less a grievance or a danger than other insects; they are simply annoying. A common way to ridicule an opponent was to compare him to a flea.[4]

In fables, plays, and jokes, fleas delivered a comeuppance to the proud and the silly, overwhelming their pretentions or simply revealing them to be inconsequential—trifling as a flea. To send someone off "with a flea in his ear," a saying adopted from the French (*la puce á l'oreille*), was a rebuff going back to the fifteenth century. Likewise, a "flea-bite," explains *A New English Dictionary*, was "anything that causes slight pain, a trifling inconvenience, or discomfort."[5] In *The Taming of the Shrew*, Petruchio berates a tailor as "Thou flea, thou nit, thou winter-cricket thou" (4.3.110). In *Henry V*, the Duke of Orleans dismisses the prowess of the English and their leader, sneering, "You may as well say that's a valiant flea that dare eat his breakfast on the lip of a lion" (3.7.1771–72).

Nevertheless, the flea did have a kind of power. It could penetrate the human body, and it was hard to kill. Despite the popular expression, the flea's strong exoskeleton made it difficult to squash, and its leaping ability let it escape attackers. Thus fleas could be both the metaphor of degradation and its instrument. An anonymous poet caught the biting power of the flea, even set against much larger animals: "I'd dare / The Lion, Panther, Tiger, the Beare / To an encounter, to be freed from these / Relentless demy-devils, cursed Fleas." The poem pinpoints the association of fleas with the devil and witchcraft, a potent connection during the European witch craze of the sixteenth and seventeenth centuries. The poet continues, "Doubtless I think each is a magic Dancer, / Bred up by some infernal Necromancer."[6] The idea had occurred to other cultures—Bohemian Gypsies, for example, thought of fleas as "Satan's horses."[7]

It was therefore absolutely necessary to subjugate fleas. Aesop in antiquity related:

> A Flea bit a Man, and bit him again, and again, till he could stand it no longer, but made a thorough search for it, and at last succeeded in catching it. Holding it between his finger and thumb, he said—or rather shouted, so angry was he— "Who are you, pray, you wretched little creature, that you make so free with my person?" The Flea, terrified, whimpered in a weak little voice, "Oh, sir! pray let me go; don't kill me! I am such a little thing that I can't do you much harm." But the Man laughed and said, "I am going to kill you now, at once: whatever is bad has got to be destroyed, no matter how slight the harm it does." The moral is "Do not waste your pity on a scamp."[8]

A victory over a flea seems a paltry thing, displaying not only the vulnerability of the flea but also the emptiness of his conqueror's bravado. What does it mean to kill a flea or, contrariwise, to be bitten by a flea? Jonathan Swift launched the insect against the hubris of his fellow poets:

So, naturalists observe, a flea
Has smaller fleas that on him prey;
And these have smaller still to bite 'em,
And so proceed *ad infinitum*.
Thus every poet, in his kind,
Is bit by him that comes behind.[9]

But Swift had more in mind than the absurdity of his competitors. Alluding to Thomas Hobbes's war of all against all, he noted, "Hobbes clearly proves, that every creature / Lives in a state of war by nature. / The greater for the smaller watch, / But meddle seldom with their match." Human beings, in other words, live in a state of constant anxiety because they know that the strongest can always be taken down by the weakest. This fear surfaced in the Bible when the young David admonishes King Saul, who is trying to kill him, "For the King of Israel has come out to seek a single flea—as if he was hunting a partridge in the hills" (1 Samuel 26:20). The reader knows that ultimately David the Flea will become David the King. Even in Scripture, the small can triumph over the great; clergy preached this lesson to centuries of biblical audiences.

Like the louse and the bedbug, the flea featured in satire directed at science, politics, religion, and society. The flea can seem the most paradoxical of insects: it should be inconsequential because of its size, but it can be trained to jump and perform—like the subjugated women and slaves it sometimes represented. And, like them, it could sometimes turn on its oppressors, causing disease.

Fleas can upend social hierarchies of master/slave, man/woman, human/ insect, and white/Black. Flea stories could be a form of social commentary, used either to demean the flea bitten or to rescue them from slavery or subservience. In a saying traced to the Roman philosopher Seneca and repeated by Benjamin Franklin, men were warned that "He that lieth down with dogs shall rise up with fleas." Fleas are both part of the human condition and a means to comment on it.

Philosophic, Scientific, and Funny Fleas

The hilarity stirred by fleas was less derisive, but more nervous, than the laughter about lice. It was more a titter than a shriek. It may be that flea stories, while often moralistic, sometimes produced less lofty lessons. The sixteenth-century French essayist Michel de Montaigne echoed many before and after: "Man is certainly mad; he cannot make a flea, and yet he will be making gods by the dozens."[10]

The charge of flea-induced idiocy goes back to the playwright Aristophanes, who in *The Clouds* accused Socrates of madness because of the philosopher's interest in fleas. A disciple supposedly asked Socrates how far a flea could jump, and how to measure the distance. Socrates solved the problem when "he melted some wax, seized the flea and dipped its two feet in the wax, which, when cooled, left them shod with true Persian slippers. These he took off and with them measured the distance." To ancient Greeks, the image of a flea wearing Persian slippers clearly showed that Socrates was ridiculous and probably insane—as well as offered a passing comment on Persians.[11]

Sadly for Socrates, this story was repeated many times in the centuries to come. In his satirical attack on all forms of authority, *In Praise of Folly*, the sixteenth-century Renaissance writer Erasmus repeated Aristophanes's story about Socrates and fleas, saying of the philosopher, "For while, as you find him in Aristophanes, philosophizing about clouds and ideas, measuring how far a flea could leap, and admiring that so small a creature as a fly should make so great a buzz, he meddled not with anything that concerned common life."[12] Erasmus, a classicist, also retailed a Latin saying: "The vain man invokes heaven if a flea bites him."[13]

Fleas symbolize the foolishness of mistaking the trivial for the important— or as the seventeenth-century poet Jean de La Fontaine remarked, a man who wants "to kill a flea would have forced heaven to lend him thunderbolt or bludgeon."[14] In fables and joke books going back to the Middle Ages, fleas signified folly, and folly sometimes endangered not only the body but also the soul.

A 1609 pamphlet depicted Democritus (535–475 BCE), the ancient philosopher and proponent of ancient atomism, telling another natural philosopher, Heraclitus, about a dream he had of a flea debating an elephant about who is nobler. At first, the elephant declines to participate because it would be beneath his dignity to contest with a mere flea. In response, the flea accuses the elephant of using his honor as a shield for cowardice. In this tale, a lower creature—a flea was very low indeed on the great chain of being—assaults not only the position but also the honor of an animal above him. The flea cries, "Vertue consists not in the quantite [the elephant is huge], but rather is an inward qualitie." Thus the tiniest animal can claim superiority to the largest.[15]

Democritus thinks the whole debate is extraordinarily funny, but Heraclitus is not amused: "To see how great ones still would greater be,/And none contented with their own degree." He bemoans the fact that bloodthirsty fleas, like newly risen men, bite the hand that feeds them:

> This is the wicked custome of our dayes.
> To seek their ruyne [ruin] who first did them raise.
> Foul sinn hath made her markes upon thy backe
> And (like her selfe) hath closed thee in black.[16]

The black flea is sin, expressed as a parasitical insect. It is demonic and thus presumably not so funny to those it bites. The elephant despises the ignoble flea, but while the elephant gets the last speech, a puritan-like meditation on the condition of man—"That for one pleasure, he hath twenty woes"—it is the flea who has made the better argument. Peter Woodhouse, the author of the piece, concludes, "It is a foolish toy I write,/And in folly most delight."[17]

Fleas are not just funny, but devilishly funny. A popular legend, repeated in *Harper's Magazine* in 1856, told of St. Dominic—the thirteenth-century founder of the Dominican order and the general of a Papal Inquisition—once having his reading disturbed by a flea, "the devil in spirit," hopping on the page he was reading. The saint, "to punish him for his diabolical impertinence . . . bound him as a marker on his page." The flea acted as a cursor, remaining in place while the saint reflected on a passage and following the saint's eyes if they traveled to another page. Consequently, the flea "never had a diabolical will of his own."[18]

Fleas can be tiny little demons, or even the damned themselves. In *Henry V*, after Falstaff's death, one of his companions recalls that Falstaff "saw a flea stuck upon Bardolph's nose, and he said it was a black soul burning in hellfire" (11.3.870–73). In the 1756 comic novel *The Life of John Buncle*, the hero, fascinated by natural and moral philosophy, vividly describes a gladiatorial combat between a flea and a louse. Magnified two feet by an extraordinary microscope and placed on either side of a box, "they both kept at a distance for near a minute, looking with great indignation at each other, and offering several times to advance. The louse did it at last in a race, and then the flea flew at him, which produced a battle as ever was fought by two wild beasts." The most amazing thing about these creatures, besides their ferocious hatred for each other, is their size. "But considering what specs or atoms of animated matter they were, it was astonishing to reflexion to behold the amazing mechanism of these two minute things, which appeared in their exertions during the fray." The battle between the atomistic vermin ends with the victory of the louse, "and the consequence of the dreadful engagement was, that the flea expired." The flea is insignificant even compared with the louse.[19]

Spectacular Fleas

Even before fleas became professional entertainers, their striking abilities had been noticed. Centuries earlier, Thomas Muffet described an Englishman named Mark who took a flea with a "little head, with a mouth not forked but strong and brawny, with a very short neck" and fastened it by means of a thin gold chain to a "Coach of Gold that was every way perfect . . . which much sets out the Artists skill, and the Fleas strength."[20]

Mark was the first of many smiths and watchmakers demonstrating their skills in miniaturization by attaching fleas to cunningly contrived carriages and carts that the insects then pulled, to the amazement and amusement of spectators. The fascination with the miniature, still a familiar emotion, contributed to the fascination with the flea. Indeed, the Oxford scholar Robert Burton (1577–1640), after seeing a watchmaker display a flea in harness to demonstrate his skills, remarked "that art as well as nature is never more wonderful, than in the smallest pieces."[21]

Fleas starred in exhibits of all kinds, showing and even demonstrating both the wonders of the natural world and human power over it. The architect Christopher Wren, a founding member of the Royal Society, presented Charles II with an illustration of a flea, which the monarch kept in his cabinet of curiosities. A note in the archives of the Society, cited by the historian of science Keith Moore, seems to indicate that something like the first flea circus was performed before the members of the society in 1743—perhaps reflecting an enduring interest in the tiny handed down from the society's early curator, Robert Hooke. The impresario was a London watchmaker named Sobieski Boverick, who displayed a mechanical carriage pulled by a flea.[22]

Even earlier, John Ray, a naturalist and one of the original members of the Royal Society, recounted that while traveling with his friend Francis Willughby in Italy in the mid-seventeenth century, they encountered fleas for sale chained by steel or silver collars around their necks. Willughby purchased one and kept it as a pet for three months, feeding it with his own blood until it died of the cold. Presumably, he displayed his flea for his friends and perhaps put it through its paces.[23]

By the eighteenth century, the Royal Society's study of insects was well known. The association of fleas and scientists was so well documented that microscopes were referred to as flea glasses. But rather than boosting the Society's reputation, its flea interests drew derisive laughter. In the 1785 *Memoirs and Adventures of a Flea*, the flea advises the French owner of a performing flea to give the flea "to Crane-Court, where I might amaze the virtuosos of the age, and be enrolled amongst the ever valuable collection of the Royal Society, during my life; and, at my decease . . . [H]e would himself (he said) immortalize my memory, and anatomize me for a skeleton, to be deposited in the British Museum."[24]

The point is to not to display the grandeur of the flea or of her enterprising master, but to link the Royal Society with such an insignificant object, making the experimenters ridiculous. An eighteenth-century poem linked science, sex, and absurdity: a flea recounts that "A Doctor from his Wife's plump Thigh, Gaping stedfast, took me up, And all my Beauties did descry, As poring thro' his Microscope."[25] Alexander Wolcott (aka Peter Pindar), the author of *The Lousiad*, left out the sex but kept the ridiculous and the Royal Society in a poem, describing the agony of its president, Sir Joseph Banks, who was supposedly distraught at the refuting of his alleged claim that fleas were tiny lobsters.[26]

Those engaged with these smallest of insects defended themselves by arguing that God or nature was revealed most, in Muffet's words, "in these that are so small and despicable, and almost nothing, what care? How great is the effect of it? How unspeakable is the perfection?"[27] This sentiment was repeated by an artist hired by the microscopist Anton van Leeuwenhoek to illustrate a treatise on fleas for the Royal Society in 1680. The artist, who had great difficulties drawing the minute parts of the flea, exclaimed, "Heavens! What wonders here in so small a creature!"[28] Leeuwenhoek showed in this work that fleas are not generated spontaneously from dirt or sweat or urine, but emerged from eggs into nymphs or maggots and then into adult fleas. But Leeuwenhoek realized it was the flea's skill in jumping that really intrigued "the curious," and so he pulled off one of his specimen's legs to inspect under the microscope, concluding, "if we reflect on this wonderful and complicated formation of joints in a Flea's leg, we shall cease wondering that it can leap to so great a height as we see; nor ask the question (which I have often had put to me) whether Fleas had wing to carry them so far and high?" Although the Dutch scientist knew that most people would mock him for investing his time "on that minute and despised creature the Flea," he also knew that "the Flea is endowed with as great perfection in its kind, as any large animal."[29] This sentiment was anticipated by Robert Hooke, who thought all creatures,

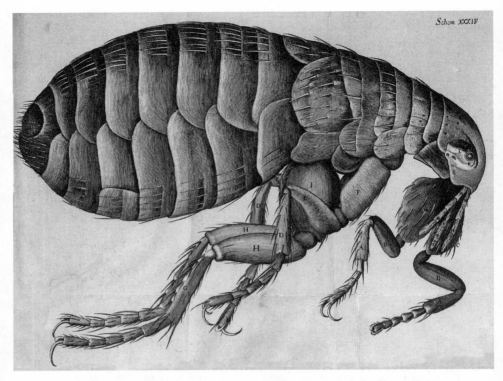

Robert Hooke, *Microscopic Image of a Flea*, engraving from *Micrographia*, 1665. Courtesy of the Wellcome Collection, https://wellcomecollection.org/works/ajveb66y

including insects, were testaments to God's creative power. In the preface to *Micrographia*, he proclaimed the wonder of observing tiny things: "my little Objects are to be compar'd to the greater and more beautiful Works of Nature, A Flea, a Mite, a Gnat, to an Horse, an Elephant, or a Lyon."[30]

Fleas inspired both the eyes and hands of observers. Two of the most famous portraits of fleas, 150 years apart, capture the nearly supernatural awe produced by close examination of these ordinary insects. The images produced by Robert Hooke in 1665 and William Blake in 1819 demonstrate the flea's enduring hold on the imagination.

Hooke's flea is famous, creeping through many books on natural history from its first appearance until today. At first glance, it seems—at least to the modern eye—like the most exact and empirical kind of observation. But Hooke's rhapsodic ode to his subject demonstrates how the microscope could elevate the most insignificant object into a wonder: "The strength and beauty of this small creature, had it no other relation at all to man, would deserve a description. For its strength, the Microscope is able to make no greater discoveries of it then the naked eye, but onely the curious contrivance of its leggs and joints, for the exerting that strength, is very plainly manifested, such as no other creature, I have yet observ'd, has any thing like it." The flea is remarkable for how its anatomy allows it to gather up its strength: "These six leggs he clitches up altogether, and when he leaps, springs them all out, and thereby exerts his whole strength at once."[31]

Science meets the language of art as Hooke continues his description: "But, as for the beauty of it, the Microscope manifests it to be all over adorn'd with a curiously polish'd suit of sable Armour, neatly jointed, and beset with multitudes of sharp pinns, shap'd almost like Porcupine's Quills, or bright conical Steel-bodkins [i.e., daggers]; the head is on either side beautify'd with a quick and round black eye."[32] Thus the flea is ennobled, a knight in shining armor ready for battle, with also the power to, in modern terms, "leap tall buildings in a single bound." But however strong and quick (a term that in the seventeenth century also denoted intelligence) Hooke's flea might have been, it is now affixed to a page by its master—the scientist as conqueror of the natural world.

But fleas could also wrest control from their observers. In contrast to Hooke's static flea, William Blake's (1757–1827) equally arresting "The Ghost of a Flea" is a nightmarish expression of a vampire beast who had appeared to Blake in a vision. Perhaps his dream was inspired by Muffet's description of the flea's "strong and brawny" tongue and its short squat neck, or maybe he had been reading *Micrographia* before going to bed.[33] A contemporary of Blake, John Thomas Smith, keeper of the prints and drawings of the British Museum, explained, "Blake said of the flea, that were that lively little fellow the size of an elephant, he was quite sure, from the calculations he had made of its wonderful strength, that he could bound from Dover to Calais in one leap."[34] But if this assessment sounds humorous, Blake's engraving is far more frightful than funny.

There is no mistaking this creature's threat and power. Blake told the artist John Varley that in his vision this species was "inhabited by the souls of such men as were by nature blood-thirsty to excess, and were therefore providentially confined to the size and form of insects; otherwise were he himself the size and form of a horse, he would depopulate a great proportion of the country."[35] To the critic G. K. Chesterton, Blake's flea incarnates the very "principle of a flea . . . bloodthirstiness, the feeding on the life of another, the

William Blake, *The Ghost of a Flea*, tempera and gold on mahogany wood, 1819. Wikimedia Commons, https://commons.wikimedia.org/wiki/File:William_Blake_-_The _Ghost_of_a_Flea_-_Google_Art_Project.jpg

fury of the parasite."[36] In Blake's obituary in the *Literary Chronicle* in 1827, "the Man Flea" is described as "indubitably the most ingenious, and able personification of a devil, or a malignant and powerful fiend, that ever emanated from the inventive pencil of a painter."[37] Here, once more, is the flea as a devil. It needs to be contained even if only by the imagination of the artist, who confines it in a curtained window with the universe pulsing outside.

Robert Hooke was a scientist and William Blake a mystic, but both demonstrated the flea's richness in meaning and significance. Both flea images possess personality and selfhood, although an observer might wonder if these ferocious creatures could possess reason or even a soul. Their eyes are alert and searching, perhaps for their next victims. Hooke's flea seems to be gathering itself for a mighty leap, while Blake's flea strides across the scene holding a bowl to collect the blood his mighty claws will draw from his prey, licking his lips in anticipation of his next meal.

These images skirt the boundary between real objects and imaginary terrors. In both pictures the fleas are distorted beyond the normal, becoming monsters and threatening the viewer. Hooke's flea is a gigantic presence in his book, occupying a foldout folio page, looming many times the size of a living flea. Blake's flea is a spirit—a ghost—but the most material and bloodthirsty ghost imaginable, even if the drawing itself is tiny, 24.1 cm × 16.2 cm. Even tinier is the image of a minute flea lying on the ground under the anthropomorphized feet of the ghost flea, perhaps a commentary on the monstrous potential of Hooke's flea, which the tiny animal resembles.

In their vivid representations, Hooke and Blake played on viewers' expectations, emphasizing the power of the artists' own perceptions of the world, whether a vision natural or supernatural. In their works, size becomes an instrument of control; Hooke's flea might be gigantic, but it is an exaggeration created by the experimenter and thus a testament to human power rather than human weakness. Likewise, Blake's terrifying flea ghost is rendered almost ridiculous by its tininess; it can only suggest the power it might possess if it was the size of a horse instead of an insect.

The fate of fleas, from the first time they were displayed by their human masters, is a sad one. From the insect's perspective, its interaction with people results in subjugation and death. When Mark chained his flea, or Hooke observed his flea under the microscope, part of the laughter and amazement the tiny performers evoked was rooted in cruelty, often a comic element we would prefer to overlook.

The interaction of laughter, cruelty, and science received explicit notice in the pages of *The Tatler*, the first journal that combined social satire and gossip, published by Richard Steele from 1709 to 1711. It contains a letter from the supposed widow of Nicholas Gimcrack, a character in Thomas Shadwell's satirical play *The Virtuoso* (1676). Another character in the play describes Gimcrack as "such a coxcomb he has studied these twenty years about the nature of lice, spiders, and insects and has been as long compiling a book of geography for the world in the moon."[38] The fictional Lady Gimcrack reports that her husband became very odd when he bought a microscope and became a member of the Royal Society. While chasing butterflies the poor man contracted a fatal illness, and while dying he told his wife "that it was a custom among the Romans for a man to give his slaves their liberty when he lay upon his death-bed. I could not imagine what this meant, until, after having a little composed himself, he ordered me to bring him a flea which he had kept for several months in a chain, with a design, as he said, to give it its manumission."[39]

Clearly, some writers in the past understood that the life of a flea was often the life of a slave. Subjugated fleas were emblematic of subjugated people, particularly women and enslaved Black people, but evoked anyone who had lost freedom and existed at the mercy of another. To flea someone—etymologically speaking—is the same as to flay them, to strip off their skin and torment them. Blake's "Ghost of a Flea" looks ready to flay, with his hooked claws and the thorn he carries in his right hand—he is demonic, and demons serve the devil. So, once again, the flea creeps into its double identity as both the subject and instrument of degradation.

Subjugating Fleas: Slaves, Gender, and Sex

In the early nineteenth century, William Blake was exceptional for his progressive and sometimes idiosyncratic views on sexuality, spiritualism, and slavery. He did the engravings for John Gabriel Stedman's *The Narrative of a Five Year Expedition against the Revolted Negroes of Surinam* (1796), a complex and conflicted view of slavery. The engravings are shocking and explicit, including a drawing of the flagellation of an almost naked female slave.

The relationship between Blake's "Ghost of a Flea" and his earlier work for Stedman seems clear. The connected words "flagellation," "flay," and "flea" all mean torture by excoriating the flesh, and all are linked to slavery. The abolitionist Granville Sharp published a letter in 1776 from a visitor to Maryland that makes the connection clear: "The punishments of the poor negroes and convicts, are beyond all conception, being entirely subject to the will of their savage and brutal masters . . . One common punishment, is to flea their backs with cow hides, or other instruments of barbarity, and then pour on hot rum, superinduced with brine or pickle, rub'd with a corn husk, in the scorching heat of the sun."[40]

Pro-slavery as well as anti-slavery advocates linked fleas and slaves. In an account published in 1791, Thomas Atwood, who had been the chief judge of the island of Dominica, gave his impressions of enslaved "Negro" peoples. His thesis was that Black people were too lazy to remove the "jiggers," or chigoes, "an insect much like a flea," from their bodies: "Many instances have been known of negros who have unfeelingly endured the pain of the jiggers, by suffering them to breed in their flesh, their feet swelled and perforated like an honey-comb, rather than be at the trouble of taking them out."[41]

If a jigger—actually a kind of tiny sand flea—was allowed to lay its eggs underneath the toe, according to Francisco de Oviedo's account of Christopher Columbus's second voyage, "a small sack the size of a lentil, and sometimes like a garbanzo" will develop.[42] It contains the eggs of the female sand flea that made the wound, keeping it open and living in it—as Amy Stewart, the author of *Wicked Bugs*, writes—"so she can receive male visitors when she's feeling amorous." She lays hundreds of eggs, which cling to the festering wound "making for a truly appalling sight."[43] The untreated foot can develop gangrene, and the unfortunate sufferer may lose toes or even the foot itself. According to Oviedo, Columbus's sailors became so desperate they cut off their own toes. Still found in tropical climates, the disease is now called tungiasis. Columbus's sailors brought back to Europe another disease from the new world: syphilis. In early modern England, its marks were sometimes called French flea bites, an appellation clearly reflecting English attitudes toward the French, venereal disease, and fleas.

The naturalist Thomas Muffet read the travel literature about fleas and slaves, including a misreading of the account of Spain's conquests in the New World by Peter Martyr d'Anghiera in *De orbe novo* (1530). Muffet mentions that "Martyr the Author of the Decads of Navigation writes, that in Perienna a Country of the Indies, the drops of sweat that fall from their slaves bodies will presently turn to fleas."[44] In this case it is enslaved Indians who suffer from this curse, but as with enslaved Black people, there is a continuum from dust to the sweat of slaves, linking both in a community of filth and corruption. A picture by the artist Augustus Earle from the early nineteenth century captures the links between slavery and fleas, showing a female slave removing a chigger, or jigger, from the foot of a white man. The victim grimaces, while others view the operation with fascination. It is clearly a kind of spectacle; the slave woman performing the extraction is both grotesque and subservient—her body is contracted into a kind of a bow, but her expertise is clear in the concentration of her gaze and the skill with which she wields the knife. The victim looks both embarrassed and ridiculous; he is a white man who has contracted a slave affliction.

From antiquity, fleas had been associated with sex; flea literature is even more sexually explicit than the most pornographic lice stories. In their stories, fleas took a more active role, and sometimes the flea was admired for its

Augustus Earle (1793–1838), *Extracting a Chigger, Brazil*, watercolor on paper, 1820–24. Slavery Images, http://slaveryimages.org/s/slaveryimages/item/2767

audacity and its intimacy with its hosts.[45] These tales ranged from funny to salacious, depending on your point of view—and your culture. Flea literature incorporated the related themes of misogyny, demonization, and enslavement, all presumably meant to produce laughter.

The most famous poem about fleas focuses on sexual conquest. In John Donne's 1631 "The Flea," the poet tries to seduce his beloved by alluding to the flea that has bitten them both:

> Mark but this flea, and mark in this,
> How little that which thou deniest me is;
> It suck'd me first, and now sucks thee,
> And in this flea our two bloods mingled be.
> Thou know'st that this cannot be said

A sin, nor shame, nor loss of maidenhead;

Yet this enjoys before it woo,

And pamper'd swells with one blood made of two;

And this, alas! is more than we would do.[46]

Here the flea swells through the verminous union of the two lovers' blood, anticipating the coital coupling the poet desires and uses his wit to achieve. The loss of her virginity, the poet argues, is inconsequential since the blood of the lovers has already been joined in a flea. Donne's object is seduction; his true feeling about fleas and women became apparent in a less famous aphorism, when he wrote, "Women are like Fleas sucking our very blood, who leave not our most retired places free from their familiarity, yet for all their fellowship will never be tamed or commanded by us."[47] In the identification of women with fleas, the insect becomes no longer the means for seduction but rather a symbol of how women can outwit and subjugate men. In "The Flea," the woman ultimately kills the flea: "Cruell and sodaine, hast thou since/Purpled thy naile, in blood of innocence?" The poet claims that the destruction of the flea is a kind of martyrdom since "This flea is you and I, and this/Our marriage bed, and marriage temple is," but the more astute reader might see a protest by the maiden against the assaults of her seducer.

The misogynistic uses of fleas are even clearer in the sixteenth-century poem "Elegia de Pulice" ("In Praise of the Flea"), at the time wrongly attributed to Ovid. John Donne clearly knew this piece of doggerel, written originally in Latin by a Dominican theologian, Petrus Gallissardus, and published in 1550. It is an example of the sardonic "In Praise of" tradition, which includes Erasmus's "In Praise of Folly" and Heinsius's "In Praise of the Louse." The author asks, "Little Flea, disagreeable pest, unfriendly to maids, how shall I sing your warlike deeds?" The poem extols the flea's power over women:

> When you further fix your sharp hidden piercer into her side, the maid is
> driven to rise up out of deep sleep. And about the lap will you wander; there, to
> you, ways are open to other members. You please yourself wherever you wish;
> nothing is hidden to you, savage. Oh! Disgusting! And I say that when the maid
> lies reclining, you ravage the thigh, and cause the gore to flow. And meanwhile
> you dared to broach even the passionate parts, and to taste the pleasures born
> in those places.[48]

The actions of the flea may be disgusting and bloody, says the author, but that does not stop him from wishing to become metamorphosed into a flea, and "wandering under the dress, up the legs, I will quickly bestir myself to those places which I choose!" If the maid scorns him when he becomes a man once again, the poet boasts, he will once again assume the form of a flea, and "Whether by Prayers, or by force I'll have her. Then she'll prefer nothing to having me as her own companion."

The flea man in "Elegia de Pulce" threatens rape; flea bites and ravishment are parallel acts, and in both the maid is subjugated by means of a flea. But somehow, in the lustful imagination of the perpetrator, she will come to love her assaulter. Similarly, in Christopher Marlowe's *Dr. Faustus*, the personified vice, Pride, proclaims,

> I am like to Ovid's flea, I can creep into every corner of a wench; sometimes like a perriwig I sit upon her brow; next, like a necklace I hang about her neck; then, like a fan of feathers, I kiss her lips; and then, turning myself to a wrought smock [i.e., an embroidered petticoat], do what I list. But fie, what a smell is here! I'll not speak a word more for a King's ransom, unless the ground be perfumed and covered with cloth of arras.[49]

Misogynistic flea fantasies always contain an element of shame and disgust. Thomas Muffet's description of the flea links female moistness, a universal trait of women according to Aristotle, with the insect: "Fleas are a vexation to all men, but especially, as the wanton poet [Ovid] hath it, to young maids, whose nimble fingers, and that are as it were clammy with moisture, they can scarce avoid."[50] Inspired by this tradition, in 1579 a group of French bon vivants had a poetry contest immortalizing the sight of a flea on the aristocratic bosom of Mademoiselle Catherine des Roches. (As Robert Burns has shown us, parasites on breasts seem to bring out the poetic impulse.) The instigator of the fun, the lawyer and wit Étienne Pasquier, added this poem to *La Puce de Madame Des-Roches*, a collection of poetry published in 1582:

> If only God permitted me
> I'd myself become a flea.
> I'd take flight immediately
> To the best spot on your neck
> Or else, in sweet larceny,
> I would suck upon your breast,
>
> . . .
>
> Or else, slowly, step by step,
> I would still farther down,

And with a wanton muzzle
I'd commit flea idolatry,
Nipping I will not say what,
Which I love far more than myself.[51]

This verse was meant to evoke laughter, but like so much flea lore, it contains violent imagery and an objectified female subject. And Catherine des Roches, a humanist poet in her own right who refused to marry, wrote a response. In her version the flea was originally "a chaste young Maiden, / Noble, wise, sweet, and beautiful," whom the god Pan wanted to marry and who asked to be changed into a flea in order to escape him. The virgin goddess Diana grants her wish: "She veiled your face / beneath a black covering, / since then, fleeing from this God, / Little one, you seek a place / Offering you a safe haven."[52] Presumably, the safe haven is Catherine's breast, thus countering the erotic poems that sought to imitate the flea's peregrinations.[53]

The English knew well the story of Catherine des Roches. John Donne's son wrote a short poem apparently commenting on it and on his father's poem: "One made a Golden Chain with lock and key, / And four and twenty links drawn by a flea, / The which a Countess in a box kept warm, / And fed it daily on a milk-white arm."[54] In *The Varietie*, a play written during the reign of Charles I by William Cavendish, Earl of Newcastle, a female character describing the skills of a serving maid declares, "she has the rarest receits to destroy moathes, and such a soveraigne medicine against fleas, that your maides need never to squeak as if they were ravis'd, peeping into their smocks before they go to bed."[55] The maids will not be violated by fleas, and by implication they will also escape other unwelcome advances.

Sex, rape, and violence are explicit in these examples of flea lore, and the association of women, sex, and fleas extended across national boundaries. The French *puce* (flea) crawls etymologically close to *pucelle*, maiden; *pucelage*, maidenhead; and *depuceler*, to deflower. With the early modern educated classes' familiarity with French, the link would have been widely known. As the art historian Crissy Bergeron has shown, the French phrase *avoir la puce à oreille*, "to have a flea in one's ear," has sexual connotations, meaning having an "amorous itch"—*oreille* means shell, and shells were used as symbols for female private parts.[56] An entire genre of slightly pornographic paintings from all over Europe are devoted to women hunting fleas in their bosoms.

In an English version of the flea hunt, by the caricaturist and satirist

Thomas Rowlandson, *An Old Maid in Search of a Flea*, September 25, 1794. Courtesy of the Trustees of the British Museum

Thomas Rowlandson (1756–1827), an old lady takes the place of nubile young woman, and the linkage of fleas and sexuality persists. The text reads:

> On record Bold Flea, with Columbus you'll stand.
> Who sought the American Shore
> Like him you undaunted explored a new Land,
> Where Man never ventured before.

This old maid is clearly hoping to lose her virginity to a flea. The iconography in this cartoon suggests that she might be a witch, with her sharp chin and sharp nose, and the cat perched over her shoulder.

Another satirical text from the late eighteenth century also links lust with a flea, and fleas with Black slavery. *Memoirs and Adventures of a Flea* (1785) is full of meditations on political and ethical themes, satirically presented by a put-upon flea arguing "that insects have nobler informed intellects than man. . . . Man, lordly man, fallen and degenerate as he is, may be taught to blush at the worth and guileless sentiment of an insect."[57]

The tale relates the tribulations of a she-flea in the form of a slave narrative. Echoing earlier stories about the peregrinations of fleas, this flea jumps across social classes. Shortly after the flea is born, she feasts on the derriere of a servant girl, who ultimately sells her to a Frenchman, the first of many people of varying rank the flea will encounter. The new owner harnesses the flea to a coach by a string around her neck and they tour the country, to the astonishment of the English, who love a spectacle. Ultimately, the flea escapes this tyranny and for a while enjoys the company of her peers in wanton abandon—she is "an errant willing female, subject to all the frailties of my sex."[58]

All the resonances of fleas, sex, and subjugation are captured in this tale, heightened by another flea's identity as a lusty female. Her partners in pleasure are described as "little black husbands," and indeed, according to the narrator, "there was not a black beau of the whole tribe, but what had a little seraglio; nor a female, but had a score of gallants at her private service."[59] Tapping into the cultural conventions that women and Black people were lustful and unruly, this account implies that they, like fleas, must be controlled—or squashed.

In early modern culture, fleas had many functions besides jumping and pulling for viewers' amusement. They often provoked a laugh, usually ridiculing human presumption and absurdity. From Shakespeare to William Blake, literature and art displayed them as little devils that represented man's capacity for sin. To science, by contrast, the intricacy of fleas' bodies demonstrated the elegance of God's architecture—which didn't keep fleas from starring roles in pornographic stories. But in observing a flea, whether through a microscope or in the illustrations of a book, the examination itself embodies subjugation and may no longer be that funny. When fleas and slaves are pressed together, humor melts into horror.

The seventeenth-century mock-heroic poem "On the Biting of a Flea" demonstrates the vulnerability of man and the power of the flea in their relationship, even if the connection is funny:

> Blood-sucking Tyrants, will you ne'er depart?
> Why do you hang in Clusters on my skin?

Come one-to-one and try what you can win.
You coward Ethiop Vermin![60]

Ultimately, off the page, across countries and continents, the flea gets the last laugh, with the revelation of its lethal secret identity—as the instrument for killing millions of human beings.

CHAPTER SEVEN

Fleas Become Killer Comedians
Literal and Linguistic Weapons

Perhaps fleas caused the French Revolution. Marie Antoinette, whose extravagant styles of dress incensed the French lower classes as much as her apocryphal remark, "Let them eat Cake," loved puce-colored gowns. In 1775, after Louis XVI complimented her dress choice, by saying "It is the color of a flea (*puce*)," she and all the fashionable members of the court started wearing it: "Every lady at Court wore a puce-colored gown, old puce, young puce, *ventres de puce* [flea's belly], *dos de puce* [flea's back], etc. [And] as the new color did not spoil easily, and was therefore less expensive than lighter tints, the fashion of puce gowns was adopted by the [Parisian] bourgeoisie."[1] In 1792, while awaiting her doom in prison, the queen again started wearing a puce-colored gown. The historian Caroline Walker suggests that the choice of color might have been utilitarian, since the tower was dirty, but it may also have been a comment on her captivity.[2] Fleas, even with their omnipresence in prison narratives, are not easily confined. As one wit put it in *Harper's Magazine* in 1859, in response to the question of whether a flea can be tamed, a flea "will leap and kick" for liberty "until the end."[3]

Fleas were linked with slavery in both pro- and anti-slavery texts in the nineteenth century, continuing the association made in earlier times. In one slave narrative, a slave recounts an instance when he was given five hundred lashes, "At the time my back, from my neck to my hams, was completely mincemeat ... I could not keep the vermin out of my flesh for weeks at a time. I was a moving stench to myself and all who came near me."[4] But it was not only African Americans who were associated with fleas. The suffragettes Susan B. Anthony and Lucy Stone were called parasites and fleas.[5] Ironically, the suffragettes' commitment to free and open dialogue at their meetings resulted in disruptive speeches by a couple of marginal characters who Ralph

Waldo Emerson described as "the fleas of the convention."[6] Wherever they appear, fleas seem to disrupt the status quo.

Maybe fleas' quest for freedom is why they have played such a big role in the entertainment industry. Their struggles under captivity fuel many of the antics they perform in flea circuses: fighting, playing instruments, waltzing. One flea observer during the height of their popularity in 1915 wrote that fleas are not trained, but "the performance can only be attributed to their desire to escape."[7] Although the audience must know on some level that the fleas are not entertaining them out of a desire for applause, the ambiguity about the impetus for the performance allows them to clap.

Fleas loom large in our imaginations because they seem both clever and inescapable. It seems we should be able to master them, but they remain impervious to our best efforts. Even when we think we're getting the upper hand, they come back to bite us—or our pets—where we're most vulnerable. Fleas are an unconquerable symbol of the power of the natural world, even in modern times, over the most vaunting claims of mere people or aristocrats.

These tiny creatures were used to point out human arrogance, pretension, and prejudice, a practice that follows them into the twentieth and twenty-first centuries. A headline in the satirical magazine *The Onion* proclaims, "U.S. Dog Owners Fear Arrival of Africanized Fleas," and the text announces:

> Panic is spreading among American dog owners, following the Center for Veterinary Medicine's Monday announcement that the arrival of a deadly mutant strain of Africanized killer fleas is imminent. "No dog is safe," CVM director Stephen Sundlof said. "While canines around the U.S. innocently fetch sticks, and chase their tails, killer fleas are migrating north at a rate of two kilometers a day. They've already invaded the border towns of Texas and California. We've got to act now, before our pets pay the price."[8]

It's clear that the ironic mutant fleas are parodic stand-ins for the Mexican immigrants who Caucasians fear will somehow bite them, just as giant imaginary fleas threaten the canine population.

Of all of the insects preying on human beings, the flea is the most widely experienced personally (except perhaps the mosquito, but that's another story). A flea can remain in stasis for a year and then emerge to suck the blood from its victim. Thus people who have just moved into a house, or travelers returning from an extended vacation, may find something unexpected awaiting them. Those mourning the death of a beloved pet may learn the less benign aspects of sharing a habitat with Fido or Phoebe, because once they lose an animal host, the fleas settle for second best—the pet owners.

Fleas' ubiquity makes them a perfect metaphor for the dangers of modern civilization. Movie theaters, places where all sorts of people encounter each other, were once referred to as flea pits. Less savory hotels are fleabags. Although flea markets have been rehabilitated into places to find a bargain or a priceless antique, originally they were simply places that offered used clothes and shabby furniture. The first flea market, the *marché aux puces* in Paris, came into existence when Napoleon III remodeled the city with broad boulevards and chased lower-level vendors to the outskirts.

In the twentieth century, fleas continued to be an ideal vessel for commentary on social change. A 1946 *New Yorker* cartoon depicted a class uprising where trained fleas strike against their employer, "Professor" Roy Heckler, the ringmaster of their flea circus; its purpose is to ridicule a statute issued by the State Labor Relations Board. The cartoon fleas hold signs demanding a union: "We want a Union—Heckler's Fleas."[9] An accompanying story describes the fleas' demands: "a close[d] shop, also higher wage to meet the rising cost of living, also time and a half for overtime, and, of course, old age pensions." Professor Heckler, the employer of the fleas, "being a reasonable man, and recognizing the extraordinary merits of the case, he permitted the fleas to hold an election" to see if they wanted to join the union of pinball attendants, shooting gallery aids and floor men employed at Hubert's Museum on Times Square.[10]

Of course, fleas are not necessarily just a metaphor. They also carry the plague. Endemic in central Asia, plague started a modern cycle in China's Yunnan province and spread in the 1890s from Hong Kong to India, where it killed millions. In 1894, the Swiss-born scientist Alexandre Yersin, associated with the Pasteur Institute, isolated the plague bacterium, and at about the same time the Japanese researcher Kitasato Shibasaburō also identified it. By 1898, a French scientist, Paul-Louis Simond, also from the Pasteur Institute, demonstrated that fleas carried the disease from rat to man. An Englishman, Charles Rothschild—one of those Rothschilds—nailed down the connection, identifying the rat flea as the particular insect spreading plague.[11]

Fleas have the misfortune of getting bubonic plague after biting an infected rat. The plague bacillus causes a clot in the flea's abdomen, causing it to regurgitate the infected blood when biting a human, who then makes the mistake of scratching his itch and sending the deadly brew into his own bloodstream. During the onslaughts of plague in the sixth and fourteenth centuries (known respectively as Justinian's Plague and the Black Death), approximately one-third to half of the population of Western Europe died of the disease. It may be that the more recent outburst didn't get as much Western press because its victims were mostly Asian—showing again that even the most devastating Third World catastrophes are barely noted until the disease somehow escapes into the West. (As evidenced by Ebola in 2014 and the coronavirus in 2020.)

The virulence of plague made fleas an alluring potential weapon during the twentieth century. The Japanese actually devised a way of dropping bubonic plague fleas on the Chinese in Manchuria. During the Korean War the North Koreans charged the Americans with using infected fleas in biological warfare. Although that accusation was probably untrue, the American military did test flea bombs in the 1950s in the somewhat humorously named "Operation Big Itch." Thus fleas have surfaced in both hot and cold wars.[12]

Biological warfare using bubonic plague was nothing new in the twentieth century. The Black Death spread after Muslim armies besieging the Genoese city of Caffa on the Black Sea in 1345 hurled corpses of plague victims over the walls. Seeking to escape the pestilence, a ship made it to Sicily, bringing the disease to Europe. None of the approximately thirty million casualties—or almost half the population of Europe—suspected that the fleas on rats (or gerbils, according to a recent scientific study) had carried the plague.[13] Modern times' superior medical knowledge enabled the weaponization of fleas, but the practice has caused almost universal revulsion: fleas may be small, but their bite can be deep.

In arguably the shortest poem in English ever written, entitled either "Fleas" or "Lines on the Antiquity of Microbes," the American humorist Strickland Gillian wrote, "Adam,/Had em." He thus summed up centuries-long debates about the role of insect parasites in Creation. It is finally settled: the little buggers have plagued us since the beginning of theological time. A recent discovery of a fossilized flea in amber adds the authority of science to this pronouncement. Whatever we do, wherever we go, the flea is hopping alongside—or riding on—us.

Entertaining Fleas

The linkage among fleas, demons, battles, and laughter that we saw in chapter 6 continued into the nineteenth century. An American Civil War story captured both the humor and terror of fleas. A group of Union soldiers were besieged by "those unpoetical insects known as fleas . . . the diminutive foes of man." They "itch like Satan," and one hand alone is not enough "to allay the irritation of our corporeality." The narrator of the story, desperate to escape the fleas' ravages, rides naked into the night, leaving his tormentors streaming behind him. But his companions still scratch with "demonic violence" and seek to drown their misery with drink, rolling around "more than a rural tragedian in the tent scene of Richard the Third." The response of the flea-free soldier to their suffering is emblematic of the conflicted hilarity fleas could provoke: "In spite of my pity for the poor fellows, I could not refrain from laughing."[14]

The humor of fleas is captured in another story from the antebellum South. "Loud peals of laughter" result when a slave boy reveals to his mistress before her guests that her baby had been screaming and fretful because "a blea (flea), a blea, missus! Da bin bite de poor chile."[15] Whether the guests were laughing at the flea, the slave, or the embarrassed lady is hard to tell from this account. Perhaps the boy was using the flea to ridicule his owner, once again demonstrating the power of the weak over the strong. Perhaps we don't get the joke because it is so strange to modern sensibilities.

But one contemporary of the fleeing soldier and the slave might not have thought this episode was so funny. The Reverend Henry Ward Beecher (1813–87), the brother of Harriet Beecher Stowe, seemed to lack a sense of humor, at least about fleas as metaphor. "Life would be a perpetual flea hunt," Rev. Beecher wrote, "if a man were obliged to run down all the innuendoes, inveracities, insinuations and suspicions which are uttered against him."[16] But his censorious words urging men not to be preoccupied with insignificant trifles (fleas) did nothing to stop the entertainment juggernaut that fleas, devils or not, had become by the nineteenth century.

Flea circuses fascinated from their inception in the nineteenth century and still intrigue us today. Performing fleas appeared all over the European continent, England, and the United States. The first public circus seems to have been the work of a famous Italian showman, Louis Bertolotto, who presented shows in the early nineteenth century in London and New York.[17] By the 1830s in America, flea catchers were hunting for fleas in the notorious Five Points section of Manhattan in order to provide flea circuses with *Pulex irritans*, the human flea, the only one of the species large enough to train to perform. No doubt Signore Bertolotto's audience, who he claimed included members of the royalty, found his exhibit of the exploits of tiny beings doing human-sized tasks tremendously funny. They danced, they functioned as the steeds for a tiny Napoleon and a minute Wellington at the Battle of Waterloo, and they pulled a coach that was itself driven by another flea carrying a whip to keep his steeds on task. Bertolotto maintained that his fleas were educated and had distinct personality traits, allowing them to be trained.[18] In reality, the fleas were chained, glued to the stage, and compelled to move by heating the apparatus. One observer noted, "I entered and saw fleas here, fleas there, fleas everywhere; no less than sixty fleas imprisoned and sentenced to hard labour for life. All of them were luckily chained, or fastened in some way or other, so that escape and subsequent feasting upon visitors was impossible."[19] Even in the nineteenth century, this kind of bondage generated a response from the tenderhearted. In 1822, reacting to Richard Martin's Act in Britain, the first animal rights bill, Bertolotto published an advertisement defending his show: "The improvements which L. Bertolotto has been able to adopt, precluded all Charges of cruelty to the Fleas, and the most rigorous Observer of Mr. Martin's Act will not find occasion to exert his humanity."[20]

One of the people most amused by the flea circus was Charles Dickens, who visited Bertolotto's flea circus as a young man. His 1837–38 *Mudfog Papers* satirized the workings of the British Association for the Advancement of Science and the Commission for Poor Relief. The members of the Mudfog Association for the Advancement of Everything discuss the problem of "Industrious Fleas." A member who had attended the show felt himself obliged to report that he had "long turned his attention to the moral and social condition of these interesting animals. . . . He had there seen many fleas, occupied certainly in various pursuits and avocations, but occupied, he was bound to add, in a manner which no man of well-regulated mind could fail to regard with sorrow and anger." He describes "one flea, reduced to the level of a beast of burden, was drawing about a miniature gig, containing a particularly small effigy of His Grace, the Duke of Wellington; while another was staggering beneath the weight of a golden model of his great adversary Napoleon Bonaparte."[21]

The fleas in this parody represent the poor who are compelled to labor in a nonproductive industry. Instead, the narrator suggests, "infant schools" should be established by the government "in which a system of virtuous education . . . and moral precepts strictly implemented." Any flea presumed to continue exhibiting himself "for hire, music, or dancing, or any species of theatrical entertainment, without a license, should be considered a vagabond, and treated accordingly." Dickens makes the point of his satire of the system of poor relief crystal clear when he remarks that this fate of the vagabond flea "placed him on a level with the rest of mankind."[22]

At the same time Dickens was writing the *Mudfog Papers*, he was also publishing *Oliver Twist*. Both works should be read in the context of the Poor Law Amendment Act, passed in 1834. The new law attempted to reform the poor laws of Great Britain along Malthusian and utilitarian ideas and essentially saw the poor as responsible for their own conditions and advocated work houses to control them. Thus when Dickens's character suggests that the fleas' "labour should be placed under the control and regulation of the state" and that "liberal premiums should be offered for the three best designs for a general almshouse," he was hoping to send the industrious fleas to the workhouse, where their work would both save them from the burdensome tasks they performed in the flea circus and compel them to perform "honest labour" for the good of society. It seems that the flea had little to gain in either scenario: he or she personifies the ultimate down-and-out in society, enslaved by the powerful.

Dickens was not the only writer who exploited the satirical potential of fleas. The entertainment industry was quick to pick up the erotic tradition of fleas. Written in the late nineteenth century, the pornographic *The Autobiography of a Flea* tells the story of a woman who becomes a willing victim. In it, an observant flea tells the story of a fourteen-year-old girl named Bella, whom the flea met while "engaged upon professional business connected with the plump white leg of a young lady." Bella is seduced serially—and with her complete cooperation—by a boy, a priest, and a monk. Both misogynistic and anti-clerical, this story inspired a pornographic film in 1976, showing that flea lore extends to both new media and new centuries.[23]

The Autobiography of a Flea was part of a genre of literature called "it-narratives" or "novels of circulation." These tales were narrated by some object—perhaps an insect, a coin, or a carriage—reflecting the merchandising and global commerce of the eighteenth and nineteenth centuries. As in *Memoirs and Adventures of a Flea*, the hero could go places that larger objects could not and therefore had a particularly close view of the changing social stations and attitudes of the time. These voyagers hop from person to person and follow their hosts into servants' halls, royal courts, many different beds, and across borders. Written for the emerging reading public by Grub Street "hacks," they did not aim at literary value or philosophical messages, although many referred to politicians and aristocrats of the age. In a sense, these books played the role of modern celebrity television shows and magazines, following the antics of the rich, famous, and the notorious with the ultimate aim of bringing them closer to the reader through ridicule. The objects in these stories circulated from place to place, just like the inexpensive books and pamphlets for which they provide the content.

Popular literature written by well-known literary figures during the nineteenth century also utilized flea motifs, including the satanic associations of the flea and the linkage of fleas and politics. In Goethe's retelling of the Faust legend, the devil Mephistopheles sings of a king "who had a big black flea, / And loved him past explaining, / As his own son were he." One of the young men the devil is entertaining cries, "Hear! Hear! A flea! D'ye rightly take the jest? I call a flea a tidy guest." The joke continues when the king makes the flea a minister, and not even the queen or his courtiers is permitted to scratch it or crack it.[24]

The political theme is also found in the fantastic tale *Master Flea* by E. T. A. Hoffman, published in German in 1822 and translated into English in 1826. The eponymous hero is the king of the fleas, who rules a constitutional monarchy. The flea king's enemies are Leeuwenhoek and the seventeenth-century naturalist Jan Swammerdam, who pretend to be flea tamers but are really scientists who put the flea on display under a huge microscope after enslaving him. *Master Flea* shows the potency of anti-experimentalism during the Romantic period, which rejected the adoration of science that had informed intellectual thought since the Scientific Revolution of the seventeenth century.[25] Master Flea explains to another character, "you must be aware that they [his flea subjects] are animated by an untamable love of freedom."

Master Flea is a political satire of European politics in the early nineteenth century, so it is not unusual to find the flea portrayed as a slave who used to rule a constitutional monarchy. There were liberal and nationalist upheavals throughout Europe during this period, culminating in the revolutions of 1848. The flea tamers in this story represent the tyrannical rulers who resist the republican wave sweeping Europe, from whom Master Flea ultimately escapes.

Unfortunately, the flea fell into servitude after deserting his people because of "a passion for the fair sex, that oversteps the bounds of decorum." He had fallen in love with a human girl: "I forgot my people, myself, and lived only in the delight of skipping about the fairest neck, the fairest bosom, and tickling the beauty with kisses."[26] Thus this story also demonstrates the legacy of the erotic flea tradition that was alive and well in the nineteenth century.

Flea lore culminated in a fable by the Danish fairy tale master Hans Christian Andersen. Published in English in 1873, "The Flea and the Professor" describes a circus entertainer whose wife leaves him; his only remembrance—"an heirloom from his wife"—is a performing flea whose skills come in part because he has consumed human blood. One of the tricks the man and the flea perform is to shoot the flea out of a tiny cannon. The two deserted males—man and flea—make "a secret pledge to each other that they would never separate. Neither of them would marry. The flea would remain a bachelor and the Professor a widower. That made them even." The flea and the man (now called a professor) travel to "a country of savages" where the inhabitants "eat Christian men." The professor knows this, but he is not afraid because "he was not much of a Christian, and the flea was not much of a man." The land is never clearly identified as Africa, although the natives also dine on "elephant eyes, and fried giraffe legs." The land's ruler, an eight-year-old girl who has usurped the throne from her father, falls "madly in love in the flea" after he shoots off his tiny cannon (perhaps she is familiar with the story of Queen Christina and the cannon). She keeps the flea "tied to the big red coral pendant which hung from the tip of her ear." She warns the flea, "Now you are a man, ruling with me, but you must do what I want you to do, or I will kill you and eat the Professor." Ultimately, the professor and the flea escape in a hot-air balloon and return to their native country, where their success as balloonists gains them wealth and respect: "They are wealthy folks now—oh most respectable folk—the flea and the Professor."[27]

A Freudian psychiatrist could have a field day with this story, but for us the tale's main significance lies in its many references to the themes we've been following. Andersen's hero is really the flea—who, if not superior to the man, is at least his equal until he is enslaved by the young sovereign, probably an African, who warns him that she will eat his human companion unless he obeys her. In this story, savage human beings are as cannibalistic as fleas, although they prefer to eat Christians. Clearly, the devil would approve of their menu choices. And sexuality pervades this tale: the flea is the legacy of the circus performer's wife, and the princess falls in love with the flea, who has metaphorically taken her maidenhead—the big, red pendant—which hangs from her ear (as discussed in chap. 6, *pucelage* [maidenhead] and *avoir la puce à oreille* [to have a flea in one's ear] have sexual meanings, which Andersen might have known). The former circus impresario calls himself a professor and is knowledgeable enough about aerodynamics to oversee the construction of a hot-air balloon that allows him and the flea to escape and return to their homeland.

Recent flea trainers would not be surprised by the scientific abilities of the professor. In fact, as early as 1834, Louis Bertolotto published a pamphlet titled *The History of the Flea with Notes and Observations*, in which he claimed to have experimentally determined that a flea lives for about two years and described the coital practices of fleas in a quasi-scientific anatomical discussion.[28] Although modern proprietors of flea circuses don't take up Signor Bertolotto's claim that fleas possess consciousness and individuality, they do empirically examine the actions of their performers. Walt Noon, a twenty-first-century circus producer, argues that fleas indeed possess certain individual traits. Noon also asked scientist David Watson to explain how fleas jump, which Watson does on his website, the Flying Turtle. It includes this humorous comment: "Without Resilin fleas couldn't jump worth a hoot. They would have to try to get on dogs by standing on their hind legs and hitchhiking. Probably wouldn't work too well. Volunteer hosting?"[29]

Walt Noon has made the flea circus into a career, selling DVDs with instructions on how to make one, along with a history of the art and demonstrations of the different tricks the fleas perform.[30] The highlight of his show is a chariot race between Ben Hur Flea and Messala Flea, accompanied by appropriate race patter. Happily, Ben Hur Flea wins.[31] Charlton Heston would be so pleased by this feat performed by his flea namesake. It shows the continuing power of the tiny to intrigue and humble us. The idea that such a small creature can perform such an amazing task—pulling and racing—holds up human pretension under as strong a magnifying glass as the one the flea tamer uses to show his stars to spectators.

A twenty-first-century renaissance of flea circuses recovered an art largely lost in the previous fifty years. Until 1957, there was a live flea show on Times Square, produced first by Swiss émigré William Heckler and then by his sons William Jr. and Leroy. Like other flea circuses before it, it was used satirically to comment on labor relations. The flea circus figures in the background of a scene in the movie *Easy Rider*, contributing to the film's tawdry ambience. Ultimately, as Times Square became increasingly the province of porn shops and X-rated movie theaters, the flea circus closed its doors.

Earlier movies had also proclaimed the magic of the flea circus. A 1926 *Our Gang* short depicts a flea circus proprietor losing his star flea—Garfield—

Ornate chariot used in race between Ben Hur Flea and Messala Flea. Courtesy of Walt Noon

when it jumps on a pet dog owned by a little African American boy. Hilarity ensues as the flea and its fellows infest a policeman, then a bride and groom and their wedding guests.[32] The distance between what the 1920s thought funny and what's funny now is as wide as the gap between the funny fleas of the eighteenth century and the grimy insects of today. We simply do not understand why wedding guests writhing in fury as they try to remove the fleas from their bodies could possibly provoke a laugh. Fleas highlight the strangeness of the past, but also its concerns and assumptions. In this case the wedding guests undone by the dirty children, and especially the Black child, symbolize the vulnerability of the upper classes to assaults from below. This film evokes the racial and class fears of the past toward slaves and beggars. The depiction of the policeman's scratching and partial disrobing as he searches for fleas becomes an assault on authority. The children also try to find some new fleas for the circus by lighting the beard of a homeless man on fire, underlining their distance from the completely dispossessed.

Such social commentary continues in the 1930s in a Laurel and Hardy film where the comedians are paid for their work at a circus with a flea circus, which naturally ends up infesting their shared bed in a truly fleabag boarding house.[33] In a 1945 film, the comedian Fred Allen starred as a flea circus trainer named Fred Floogle, who after many complications rises in status from a penny-ante entertainer to a millionaire whose daughter marries an exterminator. He retires his fleas, Albert and Mimi, who had been performing on a tightrope, doing high dives, and in the case of Mimi, performing a striptease. Honoring their career, he gives them a pension.[34]

By 1952, the portrayal of fleas on film migrated from comedy to tragedy. In *Limelight*, Charlie Chaplin plays a washed-up "tramp comedian" who dies after helping a ballerina to recover from hysterical paralysis. A long sequence shows Chaplin preforming his act with his "educated" fleas, Phyllis and Henry, whom he compels to do tricks by snapping a whip and threatening to squeeze his cast. Why should he bother to tame lions, he asks, when fleas are so readily available—at least until Phyllis takes a dive down his trousers? At the end of his performance, as he is bowing, the camera pans the audience only to reveal an empty theater and the end of his career. It turns out his fleas are imaginary; in his earlier career the entertainer had traced tumbling fleas with his finger in the air and apparently, at least when he first performed his magic, the audience members allowed their imaginations to dictate their response. The mind, he understood, can fool the eyes, at least until skepticism catches up with the performers.

Outside of a couple of cartoons, the flea circus vanished from popular culture for the next forty years, possibly owing to the spread of the vacuum cleaner and DDT. Human flea infestations were a thing of the past, but the flea circus has not been forgotten. In the movie *Jurassic Park*, John Hammond, the developer of the dinosaur park, explains to his guests (who are steadily being killed off) that his first foray into show business was an electronic flea circus, which was

> Really quite wonderful. We had a wee trapeze, a roundabout—a merry-go-round. . . . They all moved, motorized of course, but people would swear they could see the fleas. "I see the fleas, mummy! Can't you see the fleas?" Clown fleas, high wire fleas, fleas on parade. . . . But with this place, I—I wanted to give (show) them something real, something that wasn't an illusion, something they could see and (feel) touch.[35]

But in going for the real thing rather than the fake, for the actually gigantic rather than the tiny imposters, he has unleashed a peril that will rampage through four movies, with more to come. If he had understood the lesson of his flea circus, he would have seen that awe is created by what the perceiver thinks he sees—not in the tangible that can eat him.

Nowadays, any number of websites teach how to create your own circus, or for those lacking the necessary dexterity or insect immunity, they suggest hiring someone to do it for you. Yet it is increasingly difficult to find the human-focused fleas who are the best performers. One ringmaster had to send to Majorca for his actors. In entertainment terms, we may have gained the Internet, but we're losing the fleas. Even our pets are losing their companions in the pleasurable pursuit of scratching, although animal rights organizations are seeking to preserve their fur as a pesticide-free zone. Most modern-day flea circuses do not use real fleas. Jim Frank, a flea circus performer explains, "That's my bracket, the illusion bracket. People tune their shows to their own skill level. Some use gearing, clockwork, small motors. . . . But if I'm outside, any speck of dust can be a flea. A puff of breath and it moves, and they've seen the flea. There are many ways to influence kids to see an illusion."[36]

It may be that with fleas we want to be fooled and experience awe. That's part of the excitement. The flea impresario Adam Gertsacov, who creates the illusion of performing fleas in the Acme Miniature Flea Circus, explains, "The flea circus, like other sideshow and circus skills, is a direct connection to a simpler time unmediated by electronics and filled with the simple ability to wonder at the marvelous world around us. I would argue that the ability to wonder is one of the things that separate humans from beasts."[37]

But some flea showmen, like Walt Noon, insist that their fleas are the real thing. The Colombian sculptor and installation artist Maria Fernanda Cardoso has displayed her flea circus all over the world, including at the Pompidou Center in Paris and the San Francisco Exploratorium. (A performance can be viewed on YouTube.[38]) Like so many flea circus entertainers before her, Cardoso has named her fleas. There are the escape artist Harry Fleadini, and Samson and Delilah who lift weights, and Teeny and Tiny, tightrope walkers. Their monikers and abilities express their claims to humanity. But they are also animals that have to be bribed (with blood), tamed, and even whipped into performing against their natures.[39] Their subjugation tells us that nature might be controlled, at least until it gives us the plague. Fleas are a testimony to human power and human powerlessness. Thus the original flea tamer, Signor Bertolotto, noted in the nineteenth century that "The supporters of the women's rights movement will be delighted to know that my performing troupe all consists of females, as I have found males utterly worthless, excessively mulish, and altogether disinclined to work."[40]

The Science of Fleas

There's something about fleas—their strength, their tininess, their apparent ability to perform stunts—that bridges the gap between elite and popular culture. At the same time that biologists and zoologists were attempting to understand the flea's role in spreading disease and flea impresarios were continuing to exploit their skills as circus entertainers, the British banking magnate and amateur entomologist Charles Rothschild (1877–1923) became fascinated with fleas. According to his daughter Miriam, he was intrigued by "performing" fleas in a circus "and was greatly impressed by their enormous strength. He considered the feat equivalent to a man dragging two full-sized elephants round a cricket ground." Miriam recalculated this estimate according to the reckoning of modern physiologists about the strength of creatures relative to their size and argued that the flea's ability to pull is equivalent to a man pulling two sheep around a cricket field.[41]

In 1939, *Ripley's Believe It or Not* reported that Charles Rothschild paid £10,000 for a flea from a grizzly bear.[42] Whether the story is true or not, it gets its impact from the image of one of the richest men in the world collecting one of the smallest animals in the world. In fact, Charles and his brother Walter, and Charles's daughter Miriam, were all fascinated by fleas. Charles, on a collecting expedition to the Sudan in 1901, identified the oriental rat flea, which he named *Xenopsylla cheopis* in honor of the Egyptian pharaoh Cheops. Over the course of his lifetime, Charles discovered five hundred new species of fleas and subspecies of fleas, in partnership with the entomologist Karl Jordan, who cared for the collection at Tring Park, the Rothschild family home in Hertfordshire. The collection was given to the Natural History Museum in London in 1913, becoming the heart of its enormous holdings of the insect order Siphonaptera, with 260,000 flea specimens of 925 different types.[43]

Charles Rothschild was an unusual man; he encapsulated the change from dedicated amateur naturalist to professional entomologist. He hated banking and loved bug hunting, but he faithfully discharged his duties at the Rothschild bank and pursued his flea passion only outside its walls. Bouts of illness allowed him to convalesce outside of England, including his journey to Egypt and the Sudan in 1901.[44] His career as a naturalist exemplifies many aspects of the imperialistic and capitalist culture of the time. His great wealth was put to the service of empire; as an insect entrepreneur, he built a collection displaying British dominance of the world—and nature.

Collecting unusual objects was a hallmark of the Rothschild family, both in England and on the continent. One cousin specialized in *têtes de mort* (death head masks), another thimbles, and another, more conventionally, stamps. Historians see the collecting mania of the nineteenth and twentieth centuries as an example of the growth of acquisition capitalism—wealth made material. Collecting required a free market and roads, a postal service, and even a police force to get things where they needed to go. As international bankers, the Rothschilds were in a particularly good position to take advantage of the opportunities modernity provided for acquiring foreign goods— and creatures.

Charles's older brother Walter shared his interest in natural history and assembled a large collection of animals at Tring Park, including two and a quarter million butterflies and moths, a gorilla (perhaps the one cited in *Ripley's Believe It or Not*), a hundred and forty-four giant tortoises, and four zebras, which pulled his cart when he toured the estate.[45] Together, the brothers and Karl Jordan published hundreds of papers describing their discoveries.

Miriam Rothschild noted that other zoologists resented the family. She quoted her father, saying, perhaps facetiously, "Fleas are increasing anti-Semitism, as the few other students of the order think I'm too keen a competitor."[46] As with success in banking, success in classifying fleas could result in antipathy from rivals and derision from the ordinary public.

The Rothschilds' interest in fleas might have reflected their rarely acknowledged understanding that Jews, like fleas, were often characterized as parasites. In *Fleas, Flukes and Cuckoos*, published in 1952 with coauthor Theresa Clay, Miriam described the views of some biologists who "see in a parasite a form of predatory animal. Instead of killing and devouring its prey whole, it can, by virtue of its smaller size, live on the host or in it, and eat it little by little. . . . [the biologist] Elton has described the difference between a carnivorous and a parasitic mode of life as the difference between living on capital and income." In fact, she continued, "It is now quite usual to regard insects and ticks as the makers of history, the moulders of man's destiny and as one of the real enemies of the human race. It was possible to see and hear Hitler and Goebbels but it is impossible to perceive the plague bacillus spreading poison."[47]

The reference to Hitler and Goebbels in a discussion of flea parasitology evokes again the Nazi use of insects to demonize Jews, which we discussed in chapter 5. In *Mein Kampf*, Hitler proclaimed, "The Jew was only and always a parasite in the body of other peoples. . . . The Jews are a people under whose parasitism the whole of honest humanity is suffering."[48] The metaphor is expanded in the Nazi pamphlet *The Jews as World Parasites*: Jews "proliferate everywhere, acting as destructive parasitic bacteria in each host people. This destructive power became truly powerful only when deceitful commerce was combined with a religion suited to it."[49]

Hitler and his propagandists tapped into a Malthusian dialectic that saw the Aryan world as battling international Zionism and Jewish capitalism. The charge that Jews were bloodsuckers destroying the health of their hosts originated in the nineteenth century. In vermin terms, Jews were most often portrayed as lice and rats, but fleas could easily be inserted into the narrative. So perhaps when the Rothschilds collected and studied fleas, they were turning the stereotype upside down. It was not bloodsucking insects but Hitler and Goebbels who were destroying civilization. And instead of parasites destroying their hosts, Miriam Rothschild argued, they sometimes produced "either unilateral or mutual benefits." Some naturalists believed "that adaptation has here evolved beyond the parasitic relationship, with the elimination of harmful effects and a gradual substitution of mutual benefit."[50]

Some scientists acknowledged that Charles Rothschild's discovery of the rat flea benefitted humanity. The eminent epidemiologist L. Fabian Hirst declared in 1924 that "the discovery [of *X. cheopis*] is but further testimony to the essential unity of science in its bearings on the welfare of the human race, for it is the natural outcome of the purely zoological researches of Rothschild and Jordan on the systemics of the Siphonaptera."[51]

But Charles would have been the last person to claim some sort of pre-eminence. He was modest about his accomplishments, as a good English gentleman—and naturalist—was expected to be in the early twentieth century, a counter to the images of anti-Semitism. His description of finding *X. cheopis* makes his expedition sound like a genteel foxhunt. He wrote to a friend:

> I never had such a lovely time in my life. . . . [the explorer] Wollaston and I
> camped out here the Nile bank about one hundred miles north of Khartoum,
> and collected the district round about. . . . We bagged 600 mammals and birds,
> 500 fleas and a fair lot of Coleoptera and Lepidoptera. . . . five of them [fleas]
> belong to a group which in my opinion will become famous as to this group
> probably belongs the plague carriers in India.

And, as he summed up his adventure, finding the fleas was "a good thing."[52]

As taken as he was by the fleas he found in Egypt and the Sudan, Charles Rothschild was much less impressed with the people he encountered. He wrote in the same letter that recounts his collecting successes, "The faalin

Arabs are weird folk and candidly I do not like them. They SEW [his emphasis] their unmarried women folk up to ensure their keeping virtuous. They also preserve the penis of the crocodile in honey and eat it as an aphrodisiac."[53] Some fifty years later, when Karl Jordan sent his account of fleas to the US military's Department of Entomology in Cairo, his assistant had to omit negative references to Arabs and Muslims.

It is ironic that Charles Rothschild shared some of the racist attitudes he faced as a Jew. But this too reveals the cultural attitudes of the larger society. In the age of European imperialism, it wasn't unusual for the English to deplore the people they ruled. As we've seen, they felt the same way toward the Irish, the Scots, and the tribal peoples of the New World and Africa. Their attitudes toward the Indians of the subcontinent blocked their acceptance of the scientific evidence that the rat flea spread plague, even as late as 1902, after Rothschild had nailed it down. Instead, the British Plague Commission claimed that Indians caught the disease by walking barefoot over dirt and cow-dung floors. By 1908, however, after consulting with Charles Rothschild and Jordan, the commission had to acknowledge that the flea was responsible for the disease—an insect, not natives, endangered the English government of the Raj and potentially the entire world.[54]

Americans equally doubted the flea's role in transmitting the plague bacillus. After the disease arrived in Hawaii in 1899 and San Francisco in 1900, it was associated with the Chinese, who, warned the *San Francisco Examiner*, lived in filth that bred plague and put their Caucasian neighbors at risk. As with lice and typhus, fear of the immigrant other outweighed any scientific pronouncement. In 1900, fear of plague and rats resulted in the burning down of the Chinese section of Honolulu.[55]

The rat flea identified by Charles Rothschild in 1901 is known now as either the "oriental rat flea" or the "tropical rat flea." Both names emphasize the foreign origins of plague, contributing to what the linguist and cultural theorist Edward Said calls "Orientalism," a way of seeing the East (a cultural as well as a geographic construct) as the ultimate other: an exotic, backward, and often dangerous opposite to civilization. This emphasis on the social and cultural differences between East and West then becomes a justification for colonialism and imperialism and, right up to the twenty-first century, war. From this perspective, the rat flea is a cultural marker of an imagined Orient where filth, not fleas, produced plague. (It should be noted in passing that when President Trump and newscasters on Fox News refer to the coronavirus as the Wuhan or Chinese virus, they are implicitly taking part in this tradition.)

But the oppressors also live in fear of the oppressed, who might use the flea to imperil their dominance. The third plague epidemic's origin in tropical areas, and its reach as far as Australia and San Francisco, heightened the anxiety of authorities in these countries and led to a spate of books, pamphlets, and reports on the plague's nature and dissemination. Most of these discussions tried to maintain an air of scientific objectivity, but occasionally hints of underlying prejudices leaked out. So in a 1920 handbook for American military medical officers, Dr. Francis M. Munson felt obliged to explain about fleas, "Certain conditions prevailing in the tropics, such as dirty houses and people of uncleanly habits, favor the multiplication of these insects."[56]

In the early twentieth century, entomologists were nervous about alerting people to the real dangers posed by rat fleas. Even so, they hoped that allowing the general public (and especially politicians) to learn of their findings would increase their professional status and perhaps even lead to salary increases. The British zoologist Harold Russell wrote in 1913, "It is well known that the mere mention of fleas is not only considered a subject for merriment, but in some people produces, by subjective suggestion, violent irritation of the skin." But once the role of rat fleas in spreading plague had been ascertained, he continued, "The humble, but ridiculous, systematist . . . has become the benefactor of humanity."[57] The downside of all this information, however, as vented at a meeting of the International Congress of Entomology in 1948, is that scientists might be lured away from the "pure" study of insects, leaving the field to those who cared only about the economic and practical applications of their studies. Karl Jordan, in particular, worried that flea research would become too practical and lead away from the pure joy of collecting and systematizing.[58]

Many would view the entire subject of flea science as funny—the long tradition of funny fleas certainly affected their study—but flea research has implications for robotics and other fields. The flea as a key to scientific advance has redeemed the reputation of naturalists like Charles Rothschild. Opening a book on the Rothschild flea collection at the British Museum, Miriam Rothschild explained defensively, "In the past, before natural history was raised to the rank of biology and biology to the status of a learned profession, it was generally understood that entomologists were cranks. Charles Rothschild was not a crank, but he was certainly a man of unusual character."[59]

Miriam continued the family interest in entomology, producing six illustrated volumes cataloguing her father's fleas. She also figured out that the jumping mechanism of the high-stepping insect was powered by resilin, a pad of elastic-like protein located above the insect's hind legs. This insight has stood the test of time, but her explanation for how the flea launches itself through space, which provoked a heated debate in the entomological universe, has recently been disproved with a high-speed video camera. (The experiment can be viewed on YouTube.) She claimed that it was all in the insect's knees, while Hugh Bennet-Clark argued that a flea's toes propelled its jump and that any other explanation would be "silly."[60]

"Silly" is perhaps an infelicitous and even sexist view of Miriam Rothschild's contributions to science. Her own career as a naturalist showed the value of studying fleas. In 1954, the British government asked for her help with a multiplication of British rabbits causing havoc in the Australian outback. She discovered that in Britain, fleas hampered the rabbits' reproduction, but that in Australia, fleas "don't like the heat" and departed the rabbits under the scorching temperatures of the outback. She realized that Spain produced fleas that could stand the heat. An attempt to transport Spanish fleas overland was stopped at the Indian border, and then the fleas were destroyed when an animal keeper in Australia dusted her rabbits with DDT. After she returned to Spain to collect a new batch, the rabbit plague was finally contained.[61]

Lady Rothschild's strategy to save the Australian outback through the judicious application of fleas is an example of what's called economic entomology, the study of using insects to control threats to society, the state, and business. Entomologists seek to systemize species, such as the Siphonaptera, to order nature and make the study of insects rational and useful, an aim that would not be unfamiliar to Robert Hooke or Carl Linnaeus—although it might make entomologists like Karl Jordan uncomfortable in its distraction from "pure" research.

As a reward for her work, Miriam Rothschild was named a Fellow of the Royal Society. Like many exceptional women before her, Rothschild was characterized as an eccentric, a term often used to describe women doing unusual things. According to her obituary in *The Economist* in 2005, "She thought fleas beautiful. Gazing at their stained sections through the microscope, she once said, gave her a feeling as ecstatic as smoking cannabis." (One fellow scientist said Miriam had "the largest cannabis I had ever seen" in her greenhouse.) Moreover, although "A lifelong atheist, she admitted that she had been tempted to believe in a creator when she discovered that the flea had a penis."[62]

A flea vagina decorates the cover of one of Miriam Rothschild's books, and she was not alone in her fascination with flea sexuality. Many scientists have explored the complex subject of flea copulation, which involves two rods, one to penetrate the female and the other to carry the sperm. "In the business of conception," the author Brendan Lehane wrote about flea mating, "due wonderment must go to the male. He is the sexual marvel of the animal kingdom."[63] As the scientist D. A. Humphries puts it, "Male fleas have the most complicated copulatory apparatus in the insect world."[64] In Miriam Rothschild's assessment, "Any engineer looking objectively at such a fantastically impractical apparatus would bet heavily against its operational success. The astonishing fact is that it works."[65]

And it has worked for a long time. Fleas frozen in amber date to the Cretaceous epoch, 145 to 65 million years ago. The zoologist George Poiner Jr. argues that even back then fleas carried a form of the plague bacillus that may have been partly responsible for killing off the dinosaurs—fleas may be tiny, but they can downgrade an asteroid.[66] Actually, these fleas were giants—fossilized fleas found in China in 2013 were .8 inches in length as compared to .1 inch nowadays, meaning they could consume a correspondingly large amount of blood and cling to much larger animals.[67] And dinosaurs couldn't even scratch—as far as we know.

Miriam Rothschild's adventures with fleas inspired the playwright Claudia Stevens to write a performance piece starring the lady entomologist and a speaking flea. The character Miriam holds up the flea and proclaims, "Behold the noble flea, the wonderfully hardy and adept pulex irritans. . . . Circus performer, athlete par excellence (boisterous music). Jumper among jumpers! Speak to me (putting flea to ear). What's that you say?" The flea is unhappy about his performance: "The last straw was when I had to be in that flea orchestra. They tied me down, put a miniscule cello between two of my legs. Then the music box played under me and when my other legs started waving around, it looked just like I was playing the cello." Nevertheless, despite this humiliation, the flea tells Miriam how his species have survived for millennia—by trust and love and giving way to necessity.[68]

Political and Military Fleas

Miriam Rothschild was not the only person who admired the abilities of fleas. Marion Wright Edelman, the chairwoman of the Children's Defense Fund, in 2015 proclaimed, "You just have to be a flea against injustice. Enough committed fleas biting strategically can make even the biggest dog uncomfortable and transform even the biggest nation."[69]

Considering a dog as an image to be bitten for justice becomes a philosophically loaded issue for animal rights activists. In our world, fleas are slaughtered to protect the dog, a moral ranking of the value of different animals. If all living beings are sentient and can feel pain, how can we justify killing one to help the other? Norm Phelps, an animal rights activist and philosophically sophisticated thinker influenced by the teachings of the eighteenth-century philosopher Jeremy Bentham, argues that all living creatures can feel pleasure and pain, and thus experience happiness and sorrow. Phelps continues, "I think the fundamental interests of insects are identical to those of all sentient beings: Achievement of happiness, avoidance of suffering, continuation of life, avoidance of death." The calculation of what to kill, especially if the insect is dangerous to other beings, is a "Sophie's Choice" applied to the animal world, and the lesson is that unavoidable sorrow is the lot of all creatures, from humans to fleas.[70]

The morality of flea annihilation becomes even more ethically complex as a question of theology. One argument-from-design proponent contends that if we follow the rationale of those who want to protect all animals, there is no basis on which to decide whether it is better to kill one species rather than another.[71] One Hindu sect, the Jains, refuses to kill insects out of a regard for all creation, not to mention a belief in reincarnation. If a flea is Uncle John, we don't want to short-circuit his karmic journey.

Mostly, however, people are on the side of larger animals in the cycle of this life. Thus American pet owners have been jailed for allowing their pets to suffer from excessive fleas.[72] Some sellers of flea products for dogs and cats even state that flea bites can kill pets from blood loss.[73]

The struggle between dogs and fleas has become a potent metaphor in the vocabulary of conflict. The journalist Robert Taber, in his classic study of guerrilla warfare *War of the Flea*, written in 1965 during the Vietnam War, makes this analogy: "The guerrilla fights the war of the flea, and his military enemy suffers the dog's disadvantages: too much to defend; too small, ubiquitous, and agile an enemy to come to grips with. If the war continues long enough—this is the theory—the dog succumbs to exhaustion and anemia without ever having found anything on which to close his jaws or to rake with his claws."[74] Taber continues, "The flea bites, hops, and bites again, nimbly avoiding the foot that would crush him. He does not seek to kill his enemy at a blow, but to bleed him and feed on him, to plague and bedevil him, to keep him from resting and to destroy his nerve and his morale."[75]

Perhaps nothing would destroy an enemy's morale more than having fleas carrying the bubonic plague rained down on the countryside. That is precisely what the Japanese did in Manchuria during the 1930s. Ishii Shirō, the Japanese scientist in charge of the biological warfare Unit 731, oversaw the release of approximately half a billion plague fleas a year in China, killing perhaps half a million people, more than the number of dead in Hiroshima and Nagasaki. Ishii, who experimented on humans, including American prisoners, realized that he needed a kind of bomb that would not kill the enclosed fleas. Ishii, with about twenty thousand other scientists and support personnel, adapted a ceramic bomb, the type 50 Uji, to become the messenger of death. One bomb could carry as many as thirty thousand plague fleas. They were dropped on targets in Manchuria, which continued to have plague outbreaks until 1947. Building on plague, Ishii also experimented with anthrax, cholera, and typhus.[76]

By the end of the war, the Japanese were planning to use fleas against American airfields in the Pacific, and they even outfitted submarines with biological weapons to attack San Diego. Fortunately, the ship carrying the weaponized fleas was sunk on its way to a staging area, and General Umezu Yoshijiro, chief of the Japanese Imperial General Staff, called off the submarine attack on March 26, 1945, stating that "if bacteriological warfare is conducted, it will grow from the dimension of war between Japan and America to an endless battle of humanity against bacteria. Japan will earn the derision of the world."[77]

By 1945, the Allies were also engaged in biological warfare research—the Canadians led this effort to use botulism toxins and anthrax as weapons. But the Western leaders decided that they would only use such weapons in retaliation against countries that employed them first. Nevertheless, particularly as the Cold War heightened tensions between the United States and the Soviet Union, the military and secret service agencies became intrigued by these new methods of waging total war. The Americans extended de facto immunity to Ishii Shirō, probably at the command of the highest members of the US government and military, including Douglas MacArthur. In 1947, Ishii gave the United States the records of his experiments, and it was rumored that he visited the American chemical warfare center at Fort Detrick, Maryland.[78] Whether the American military actually adapted the Japanese program or not is open to question, but in March 1952, North Korea, supported by the Chinese, charged that the Americans were dropping voles infested with plague fleas on its territory, as well as other insects carrying cholera and anthrax.[79] A leftist organization based in New York, the International Association of Democratic Lawyers, claimed to have investigated the charges and issued a report concluding that the Americans had spread flies and other insects in Korea. These "terrorist methods" targeted the army, refugees, and civilian populations, "a most grave and horrible crime." Moreover, "the American forces are guilty of the crime of Genocide as defined by the Geneva Convention of 1948."[80]

A charge of genocide was not something to be taken lightly in the years following the Holocaust—it was the ultimate horrific crime. The commander of United Nations forces in Korea responded quickly and emphatically: "These charges are evidently designed to conceal the Communists' inability to cope with the spread of epidemics which occur annually throughout China and North Korea and to care properly for the many victims."[81] These accusations were also refuted by American and British scientists, and a Red Cross commission under the auspices of the United Nations declared them impossible to verify.[82] But North Korea was not ready to let go, and any number of left-leaning newspapers in the West were ready to believe that the imperialists were using this tactic. The World Peace Council, a pro-Soviet European organization, authorized another commission: the International Scientific Commission for Investigating the Facts Concerning Biological Warfare in Korea and China. It was led by the Cambridge biochemist Joseph Needham, a left-leaning intellectual who had lived in China during the war and had reported that the Japanese were using plague against the Chinese.[83] He was considered perhaps the foremost English authority on China. He later published a magisterial fourteen-volume account of the history of Chinese science.

Seldom has such a major figure been destroyed by an onslaught of fleas. During a visit to China in 1952, Needham was convinced by leading Chinese scientist friends that the reports of America releasing voles infested with plague fleas were true. He and his fellow commission members issued a 665-page report arguing that plague, a disease unknown in Korea for the previous five hundred years, could be spread only by human fleas, which must have been brought in by American planes. Needham believed that the Americans had learned the mechanism of flea dispersal from the Japanese and credited North Korean reports that fleas in large numbers had been found in the dead of winter (not the season when fleas are active normally) after low-flying runs by American planes. He did no investigation himself but accepted what Chinese scientists told him. His report concluded by accepting the charge that flies and other bacterial agents—that is, fleas—were used by the United States to spread disease in Korea and targeted both the Korean Army and civilian populations, "a most grave and horrible crime."[84]

It appears now that Needham fell for disinformation originating in the Soviet Union; it took years for his reputation to recover. But the United States was investigating biological warfare at Fort Detrick by 1952. In 1954 the military conducted Operation Big Itch in the Utah desert to see how best to implement flea dispersal from airplanes. One of the first attempts ended badly when the fleas escaped on board the plane and bit the pilot, copilot, and a military observer. Fortunately, these fleas were not infected with plague. Further experiments, with the fleas infesting caged guinea pigs on the ground, were considered successful. But ultimately fleas were considered too difficult to control, and the focus of entomological warfare shifted to yellow fever–carrying mosquitoes.[85]

By 1972, the antipathy toward biological warfare had become so pronounced that a treaty banning its use, with the prodigious name "Convention on the Prohibition of the Development, Production and Stockpiling of Bacteriological (Biologic) and Toxin Weapons and on Their Destruction (BTWC)," was endorsed by 182 nations, near-unanimous support. The preamble declared that the signatory states were "Determined for the sake of all mankind, to exclude completely the possibility of bacteriological (biological) agents and toxins being used as weapons," and were "Convinced that such use would be repugnant to the conscience of mankind and that no effort should be spared to minimize this risk."[86]

It seems that fleas as messengers of death have been prohibited, although there was an exception about research into defensive means to counteract biological warfare. The sweeping support for the convention is in stark contrast to the failure of the 1996 Comprehensive Nuclear-Test-Ban Treaty to be ratified by China, Egypt, India, Iran, Israel, North Korea, Pakistan, and the United States, many of the key nuclear powers. Clearly, biological and poison warfare makes many nations—and people—particularly anxious. The International Red Cross caught this apprehension in its description of the 1972 treaty: "The misuse of science or of scientific achievements to create weapons that poison and spread disease has always provoked alarm and abhorrence in the public mind," so the BTWC "was a major step towards the total elimination of these abhorrent weapons."[87]

Why should biological warfare be considered so much more terrible than other ways of killing people during war? Part of the answer, no doubt, is that such weapons can get out of control and turn against their users. They might also stigmatize users. Japan's decision not to launch plague during World War II reflected such fear. "General Umezu Yoshijiro found his moral courage," argues historian Jeffrey Lockwood, "by tapping into a sense of cultural shame."[88] But part of the revulsion must come from the delivery system and a horror evoked by entomological warfare. We no longer fear bubonic plague because of the success of antibiotics, but other insect and animal vectors remain a source of terror. Tularemia is spread by tick bites; the less dangerous but more common Lyme disease also results from a deer tick bite. Malaria, of course, is spread by mosquitoes, as is yellow fever. And the Zika virus, also courtesy of mosquitoes, seems especially frightful as it gains a foothold in the developed world. In other words, tiny things, which we should be able to dominate, can kill or sicken us. Once again, it's the revenge of the small over the large—the vulnerability of humans to forces they can barely see.

The horror of biological and particularly entomological warfare is particularly intense when deployed by the state. In the past, servants could take revenge on their masters by putting bedbugs in their beds, or convicts might shoot lice at bystanders, but a nation can massively kill with disease. The Native Americans learned this when the British spread smallpox through contaminated blankets. The historian Daniel Barenblatt, in his study of the Japanese use of biological weapons in China, concluded that "The microbe became an instrument of imperial rule. Comparisons with the genocide of Japan's ally and ideological brother, Nazi Germany, are entirely appropriate."[89] In 2020, the charge that the coronavirus may have been manufactured in a Chinese laboratory reflects this terror of microbial weapons.

Most people thinking about flea attacks do not imagine Japanese porcelain bombs, but rather the noxious sprays we use to rid our homes of flea infestations, devices promising to protect not only our pets but also ourselves. Anyone who has suffered from flea bites would do just about anything to save his skin from being invaded and corrupted by fleas (as by bedbugs or lice). So we use flea bombs to attack our house with neurotoxins, which we would not otherwise tolerate.

Likewise, we want to protect our beloved pets from the scourge of fleas. Flea collars and flea baths are a multibillion-dollar industry, but these products carry their own serious byproducts. The chemicals in some flea collars, with the scary names of tetrachlorvinphos and propoxur, are particularly dangerous to the health of children. According to the Natural Resources Defense Council, "these chemicals cause a variety of poisoning symptoms," including "nausea, vomiting, diarrhea, wheezing, and tearing eyes" and can potentially kill pets and even human beings.[90] Instead of Fido being saved from an onslaught of fleas, he unwittingly may become part of a disease vector. It would be another victory of the humble flea over humankind.

Admittedly, fleas may not generate the kind of antipathy set off by lice or bedbugs. How we view the flea is a question of perception, both of the insect and ourselves. They are tiny commentators on the comedy and tragedy of life. Even though they can carry gruesome disease and destruction, they still seem harmless or simply annoying. Fleas have been used for the most nefarious of purposes—to imperialize and wage war. But they can also operate as avatars for those striving against the powerful and pretentious. The bassist for the Red Hot Chili Peppers, who recently called Donald Trump a "silly reality-show bozo and blustering guy who likes getting attention," is professionally known as Flea.[91]

And the name of the sexual adventurer Fleabag, in the eponymously named television show *Fleabag*, which won several Emmys in 2019, was chosen by its writer and star Phoebe Waller-Bridge because she "wanted something that would create an immediate subtext for the character. So, calling her 'Fleabag,' calling the show *Fleabag*, gives the subtext of Fleabaggy-ness."[92] Apparently, we can rely on fleas to provide the metaphor for sex, sordid entertainment, and dirt in the twenty-first century as well as in the past. When it comes to fleas, the phrases "it's history" and "in the day" do not apply.

The progression of fleas from humorous sexual adventurers in literature to exemplars of sexual virtuosity in modern scientific accounts indicates how pervasively this tiny insect continues to intrigue. Once the flea's role in spreading plague was understood, its ability to propagate became not only amazing but also dangerous. Fleas become the harbingers of the darker side of nature, which can come back to bite us in the backside. It becomes increasingly necessary to reduce the flea's status by taking advantage of its stature, to bring the flea back as sideshow entertainers who can be tamed and taught. Hence the reinvigoration of flea circuses, making people gasp with laugher rather than fear. A recent cartoon catches the ambivalence fleas continue to engender.

The seeming insignificance of fleas did not prevent scientists from observing them, artists from depicting them, or showmen from capitalizing on them. The foremost modern chronicler of flea lore, the historian Brendan Lehane, writes, "The odd thing is that somewhere in man's inscrutable

You see, Fleas don't have many career choices
open to them: I can either join the Circus
or spend my life as a parasite...

Ralph Hagen, "You See, Fleas Don't Have Many Career Choices." Courtesy of Cartoon Stock, www.cartoonstock.com

bosom there does lurk a soft spot for fleas."[93] Somehow, even though we now know how dangerous the flea can be, the insect retains its role as the invertebrate comedian, whether or not we are laughing through our tears.

Until we scratch, we can only smile.

CHAPTER EIGHT

Rodents Gnaw through the Centuries

In 1797, the moralist and evangelist Hannah More described one of the less desirable members of society, a thief and ratcatcher named Black Giles:

> Among the many trades which Giles professed, he sometimes practised that of a rat catcher; but he was addicted to so many tricks that he never followed the same trade long. Whenever he was sent for to a farm-house, his custom was to kill a few of the old rats, always taking care to leave a little stock of young ones alive sufficient to keep up the breed; for, said he, "If I were to be such a fool as to clear a house or a barn at once, how would my trade be carried on?" And where any barn was overstocked, he used to borrow a few from thence just to people a neighbouring granary which had none; and he might have gone on till now, had he not unluckily been caught one evening emptying his cage of young rats under Parson Wilson's barn-door.[1]

Black Giles encapsulates all the qualities of the rats he pretends to pursue: he is sly as well as a thief and a breaker of boundaries. The underlying theme of every rat narrative in premodern Europe and America is that rats are dangerous, and that rat-like people share rats' power to destroy or pervert man and nature. In *Macbeth*, the First Witch promises, "And, like a rat without a tail, I'll do, I'll do, and I'll do."[2] What she will do is "Maleficia," or cause harm or death to property or people. In *King Lear*, Edgar, feigning madness, claims that when he is under the grip of "the foul fiend," he eats "mice and rats, and such small deer, / Have been Tom's food for seven long year."[3] Such a diet is clearly unnatural. In the King James Bible, Leviticus enjoins, "These also *shall be* unclean unto you among the creeping things that creep upon the earth; the weasel, and the mouse, and the tortoise after his kind" (11:29). Mice and rats were considered the same species in King James's England, so this prohibition told believers that eating these creatures was not only disgusting but also heretical.[4]

The dangerous qualities of rats and their hunters were often portrayed in legends and folktales. Not surprisingly, when the witch craze raged through Europe in the sixteenth and early seventeenth century, rats were linked with the devil, just like lice and fleas. Religious leaders did not hesitate to claim that Catholics or Protestants or Anabaptists worshipped Satan and that their demon familiars appeared in the shape of rats. The medieval story of the Pied Piper, who rid the German town of Hamelin of its rats—and, after the city leaders refused to pay him, of its children—spread quickly throughout Europe and was told in England by the beginning of the seventeenth century.[5]

Whether rats are associated with the supernatural, with witchcraft, with folklore (as in the Pied Piper legend), with religion or politics, or with social upheaval, they represent a force out of control. Alternatively, they are controllable only by a power that is more than human—Satan or God, witch or angel—or by a power that seems less than human—the ratcatcher. Rats are the ultimate other; they signify a part of nature that is unnatural and can turn on nature itself, even to the extent of cannibalizing their own species.

We know now that one of the most calamitous events in human history—the bubonic plague—was the gift of flea-infected rats to humankind, but this knowledge came only at the very end of the nineteenth century.[6] In a practical sense, during early modern times rats were despised as major competitors for human food resources. The historian Mary Fissell highlights this aspect of the rat–human relationship. Rather than responding with disgust to the filthy habits of rats, which "perhaps . . . is something of a luxury" only possible in recent times, earlier people considered rats and other creatures to be verminous because they "poached human food, often items which were ready for human consumption; vermin ate things in which humans had already invested considerable time and effort."[7] Edward Topsell, a seventeenth-century naturalist and one of the few writers who distinguished between rats and mice, complained that rats "are more noysome [noisome] then the little Mouse, for they live by stelth, and feed upon the same meat that they feede upon, and therefore as they exceed in quantity [i.e., they are bigger than mice], so they devoure more, and doe farre more harme."[8]

In their efforts to purloin human food, rats are stealthy and extremely difficult to catch.[9] The cunning of rats is a truism—and perhaps a truth—going back to antiquity and confirmed in modern laboratory studies. Francis Bacon, the father of empirical science, stated the common belief, "It is the Wisedome of Rats, that will be sure to leave a House somewhat before it fall." Legendarily, rats will also desert a sinking ship, creating an adage often applied to humans as well. Thus many politicians and courtiers, throughout early modern times, were viewed as detestable rats who put their own interests first. This analogy crossed the ocean to the United States; in early America the *Vermont Republican* declared: "It is a maxim among sailors that before the vessel is to be lost the Rats will desert her. There has been a wonderful desertion of Rats lately from the Federal Ship."[10]

Hatred of rats in England increased when the brown rat (*Mus norvegicus*) displaced the black rat (*Rattus rattus*) in the eighteenth century, at almost the same time as the perceived arrival of the bedbug. The eighteenth-century writer Oliver Goldsmith, after describing the blood struggle between native black rats and invading brown rats, explained, "The Norway rat has the same disposition to injure us [as the black rat], with much greater powers of mischief. . . . But nothing that can be eaten seems to escape its rapacity." Fortunately, these rats "eat and destroy each other," thus limiting their "amazing propagation" and preventing their ravages from destroying humanity.[11] Goldsmith's source for this claim was one of the most important naturalists of the nineteenth century, Georges-Louis Leclerc, Comte de Buffon, whose description of rat cannibalism is downright terrifying: "When a famine is created by too great a number being crowded into one place, the strong kill the weak, open their heads, and first eat the brain, and then the rest of the body. Next day, the war is renewed, and continues in the same manner till most of them are destroyed."[12]

The English feared that they were under attack by the natural world; rats seemed to undermine traditional hierarchies in politics, society, and nature. Some naturalists thought that the brown rat arrived in ships from Norway in the 1730s, but this species was also referred to as the Hanoverian rat, linked with the despised Hanover dynasty, which had taken the British throne in 1714. Just as their new monarchs were considered foreigners who despoiled the country, their eponymous rodents were eating the British Isles out of house and home. Sometime between 1750 and 1770, the brown rats arrived in America, shipping out in British ships with their fellow colonists.[13]

In earlier times, even the less threatening black rat was associated with those who were undermining the political and religious status quo; three of Richard III's hated councilors, William Catesby, Richard Ratcliffe, and Francis Lovell (whose heraldic symbol was a wolfhound), were attacked by a me-

Jean Leon Gerome Ferris (1863–1930), *The Rat-Catcher*, ca. 1878. Courtesy of the Pennsylvania Academy of the Fine Arts, Philadelphia, John S. Phillips Collection

dieval punster, William Collingbourne, as "The Catte, the Ratte, and Lovell our dogge." This was not meant to be complimentary.

The only way to control rats was through ratcatchers, but these destroyers of rodents were often considered dangerous themselves. They shared the characteristics of their prey. Like rats, they undermined hierarchies. They roamed from place to place and were therefore outsiders, even if they spoke English. Instead of running from rats, they seemed to embrace them—their depictions reflected their intimacy with the enemy. In this nineteenth-century drawing, the ratcatcher is plainly a vagabond who alerts his clients to his business by having live rats swarming around his body. He seems to be looking around warily, perhaps anticipating the antipathy his profession arouses, even if his actions helped his customers.

Rats were omnipresent in the fields and towns of Europe, and efforts to control them were prominent in society, religion, politics, and literature. Like other vermin, rats swarmed across boundaries, seeming to enjoy living with human beings, eating their crops, invading their homes and biting their children. Next to domesticated animals—often used to hunt and destroy them—rats were the mammals most familiar to people. When natives of farms and towns moved into British cities, the rats moved with them, exchanging hayricks for sewers and basements.

Fabulous Rats

Rats had a particularly bad reputation in early modern England. In a 1608 play by John Day, a husband decides to have his wife, a "she rat," killed because of her supposed adultery; moreover, a friends urges him to kill her with "rats bane," a form of arsenic because "This she Rat is a Devill."[14] It all turns out well in the end—it is a comedy—but the symbolic resonances of "rat" remain clear.

The lascivious nature of rats made them natural allies of the devil and his minions on earth, the witches. Both learned and common people believed that witches had sexual relations with demons who took the forms of different animals, including rats. In the play *The Witch of Edmonton* (1621), the so-called witch, an old lady named Elizabeth Sawyer, sells her soul to the devil to get revenge on her neighbors. She proclaims,

> I have heard old Beldames
> Talk of Familiars in the shape of Mice,
> Rats, Ferrets, Weasels, and I wot not,
> That have appear'd, and suck'd, some say, their blood.

The creatures that suck the women's blood are penetrating the bodies of the witches and essentially having sex with them. Like them, the witch of Edmonton will chose evil:

> Abjure all goodness: be at hate with prayer;
> And study Curses, Imprecations,

Blasphemous speeches, Oaths, detested Oaths,

Or any thing that's ill; so I might work

Revenge upon this Miser, this black Cur.[15]

The play *The Witch of Edmonton*, written in 1621 and published in 1658, was based on a true story, the arrest and execution of Elizabeth Sawyer for witchcraft in 1621. In her confession at the trial, Sawyer admitted that the devil came to her and sucked her just above her fundament, or backside, from a teat. Whether the devil takes the form of a rat in the actual act is not mentioned in the confession, but another confession by another accused witch is more precise. According to Philippa Flowers at her trial,

shee hath a Spirit sucking on her in the forme of a white Rat, which keepeth her left breast, and hath so done for three or foure yeares, and concerning the agreement betwixt her Spirit and her selfe, she confesseth and saith, that when it came first vnto her, shee gaue her Soule to it, and it promised to doe her good, and cause Thomas Simpson to love her, if shee would suffer it to sucke her.[16]

Philippa lusted after Thomas Simpson, who "presumed to say, that she had bewitched him; for he had no power to leave her" and the devil lusted after her, coming to her in the form of a white rat. But Philippa was not the only one in her family who gave her soul to the devil. Her sister Margaret and her mother Joan also agreed to worship him when he came to them in the "pretty" forms of a rat or a dog or a toad, and their compacts "were ratified with abominable kisses, and an odious sacrifice of blood."[17] Another friend of theirs, Joan Willimot, also had a rat familiar. Collectively, these women are known as the Witches of Belvoir.

Diabolical rats and other familiars gave women power, if only in the minds of their neighbors and employers. Joan Flowers was so ugly and abusive her neighbors believed she was a witch even before the trial began: "yea some neighbours dared to affirme that shee dealt with familiar spirits, and terrified them all with curses and threatning of reuenge." But the major cause of the charge of witchcraft was because Sir Francis Manners, Sixth Earl of Rutland, and his wife believed that the Flowers had caused the deaths of their male children and the infertility of the wife.[18]

Historians have proposed many reasons for the persecution of witches, including the belief that women—and especially old women—were sexually voracious and the fear that women might undo the traditional patriarchy. Some women were marginalized in the village economy because they were widows or spinsters living in poverty who may have supplemented their meagre earnings with love or healing potions or as midwives.[19] The charge of witchcraft often reflected the notion that women had subverted their maternal role, nurturing "demonic imps" rather than nourishing their own children.[20] But these interpretations do not explain why rats or mice or weasels or dogs or cats were associated with witchcraft. These animals share a taste for human food and cross lines to get it. They suck nourishment from the provender that is supposed to nourish man and beast, just as witches allow their familiars to suck their blood and body, subverting nursing into a diabolical activity. Rats especially have a reputation for unbounded lust and procreation. Mice, rats, and weasels were all considered verminous, and by association, the creatures that hunted them—cats and dogs—could also be considered unclean.[21]

One belief in early modern Britain reflected the threat rodents posed to the poor, as well the connection between rodents and sex. According to Topsell, "in general all Mice, and not only the white Mouse, are most desirous of copulation. And when they are in copulation, they embrace with their tails, filling one another without all delay. By tasting of Salt, they are made very fruitful . . . by the licking of Salt, do ingender and conceive with young without any other copulation."[22] The power of salt in conception was a belief going back to antiquity and may be associated with the fact that semen and urine are salty.[23]

Moreover, the linking of rodent procreation with mice's long tails—rats have even longer ones—clearly has phallic meaning. The nursery rhyme "Three Blind Mice" reflects this association, although somewhat obliquely. The first appearance of the rhyme was in a collection of songs published by Thomas Ravenscroft in 1609: "Three blinde Mice, three blinde Mice, Dame Iulian, Dame Iulian, the Miller and his merry olde Wife, shee scrapte her tripe [and] licke thou the knife."[24] In medieval and early modern England, millers were often associated with lust, as Chaucer's *Miller's Tale* famously demonstrates, and they were often accused of stealing wheat or taking an unfair amount as their payment.[25] In this way, they were like rats and mice. Old women were considered especially lustful, as we saw in the accusations of witchcraft. In this iteration, the merry old wife is scraping the tripe, a part of the animal associated with the buttocks, a part of the devil's body that the witch was often accused of kissing. Then, in a change of subject, we discover that mice have been blinded because they licked her knife, yet another phallic refer-

ence. Presumably, a seventeenth-century audience would have understood
the song's sexual connotations.

Another ballad, "The famous Ratketcher, with his travels into France, and
of his return to London," also played on the powers of rats. The ratcatcher is
"The soundest blade of all his trade" who "Upon a Poale he carried/Full
fourty fulsome Vermine." The ratcatcher is well endowed with blades and
poles to kill rats but "Whose cursed lives without any Knives,/To take he did
determine." Instead of using knives, he kills his vermin with arsenic and poppies, a drug taught to him by an African, and other herbs he learned in India.
This master of rats is somehow foreign and dangerous, especially to women.
In London, he so lusts after a maiden that he baits her with a food that "would
kill no Rats nor Mice"; the bait is clearly his penis:

> And on the Baite she nibled
> So pleasing in her taste,
> She lickt so long, that the Pyson [poison] strong,
> Did make her swell I'th waste [waist].[26]

The "waste" that swells should be read as "waist"—the lecherous rat-
catcher has left the woman pregnant. He flees to the countryside to avoid
responsibility, carouses with rogues and Gypsies, and is such a prodigious
drinker "that it was doubtful whether,/He taught the Rats, or the Rats taught
him/To be druncke as Rats, together." Unfortunately, his adventures leave
him with a sore "bag" and "flag," that is, scrotum and penis. On repairing to
France, he consults another ratcatcher about his "fiery burning" and just "as
Witches common,/must use anothers ayding [aiding]," the Frenchman (the
French are experts on the French disease, or syphilis) gives him a potion to
cure his problem. Home he goes to England, where he finds "An Ugly Wench
... whose Nose was knawne [gnawed] with Vermin." Whether this syphilitic
woman is the one responsible for his disease is left unclear, but in the rest
of the ballad, the ratcatcher continues on his merry way, refusing to aid any
maiden or woman unless she sleeps with him.

Samuel Pepys owned a copy of "The Famous Ratketcher," which was probably first written down at the beginning of the seventeenth century. We can
imagine the famous, and famously bawdy, diarist guffawing over its lyrics.
Pepys belonged to the Royal Society, whose members were not as amused by
the antics associated with rats. Robert Boyle (1627–91), next to Isaac Newton
the most important member of the Society in the seventeenth century, re-

counts the experience of "a Gentleman, a strong and resolute Man, who had been long a Souldier" who "was strangely fearful of Rats, and could not endure the sight of them." This gentleman, who had been ill for a long time and traveled extensively to find a cure, "coming at length accidently and suddenly into a place where a great Rat was in a corner . . . he furiously leap'd upon him . . . and thereby put him into a fright which freed him from the Ague [i.e., illness]."[27] The idea that fright could cure a disease seems no more scientific than the belief in witchcraft, but Boyle credited its possibility because a gentleman he knew testified to experiencing this cure. Boyle felt that "witnessing" by reliable observers established truth.[28] So, at least indirectly, rats could help effect a medical cure.

Boyle believed that it was necessary to dissect animals, including rats and mice, to understand nature and God's plan for the natural world. Unlike René Descartes, who maintained that animals were automata, like machines, and could not experience feelings, Boyle acknowledged that animals could feel pain.[29] The fabulist Jean de la Fontaine (1621–95) took issue with Descartes when he published his version of Aesop's *Fables* between 1668 and 1694. Although his work was not translated into English at this time, most educated Englishmen could read French. In "The Two Rats, the Fox, and the Egg," La Fontaine argued that while animals may not possess the same kind of rationality as men, they still think and feel. To prove this, he told of two rats who find an egg and in order to protect their meal from an encroaching fox, "one of them lay upon his back and took the egg safely between his arms whilst the other, in spite of sundry shocks and a few slips, dragged him home by the tail." And so, he concluded, "After this recital, let any one who dare maintain that animals have no power of reason."[30]

La Fontaine's endorsement of animal rationality was based on the great chain of being: "We must allow to the beast a higher plane than that of plants, notwithstanding the fact that plants breathe."[31] The fabulist was quite serious about this argument, reflecting hundreds of years of French, German, and Italian legal history, treating rats as animals who could understand enough human language to obey an order to appear in an ecclesiastical court on a charge of destroying fields and to abandon those fields when so ordered. Historians and anthropologists have argued about the meaning of animal trials since E. P. Evans first wrote about them in 1906; he saw them as "magical hocus-pocus" and held "in the interest of ecclesiastical dignities to keep up this parody and perversion . . . since it strengthened their influence and extended their authority by subjecting even the caterpillar and the canker-worm to their dominion and control."[32] Most recent writers dispute this univocal—and prejudicial—explanation of this practice and emphasize the role of trials in establishing a sense of order and control in a population trying to understand disasters caused by hungry vermin. As is often the case in the anthropomorphizing of animals, their resemblance to human beings is not simply metaphorical but even literal—at least to those making the analogy.

Animal sentience and rationality were recognized as part of the order of nature both in England and on the continent. The English domestic guide *The Vermin Killer* (1680) went a step further, by assuming that rats and mice are naturally altruistic. The author advises those who want to get rid of the vermin to heat water in an earthen pot and throw two or three live rats or mice into it, "and all the Rats and Mice in the house, hearing the Cry of those in the Pot will run immediately to the Place . . . as if they intend by force to deliver the Rats and Mice in the Pot." He also suggests putting the dregs of oil in a brass or copper pot, placing it in the middle of the room, and "all the Rats and Mice will make their Appearance, as if it were to be an Assembly of an Army of Rats and Mice." Why a copper or brass pot does the trick is not explained—perhaps it's a reference to the cauldrons used by witches—but once the rats are gathered, either by pot or sympathy, they can be exterminated.[33]

Another method for killing rodents in *The Vermin Killer* speaks directly to the witch connection: "Take the Head of a Rat or Mouse, pull the Skin from it, and carry the Head where the Mice and Rats usually come, and they will immediately be gone from thence, Running altogether as if they were bewitched, and come no more."[34] Mary Fissell comments on this advice, "This display of the tiny head reminds one of the contemporaneous human executions for treason, when Londoners routinely saw the heads of the executed stuck on pikes as dreadful warnings."[35] When rats are bewitched, they become like witches; they are enemies to God and therefore deserve death by execution.

The reference to assemblies and armies in *The Vermin Killer* also implies a political dimension to rat incursions. By 1680, when *The Vermin Killer* was first published, the English feared the accession of the Duke of York, later James II, to the English throne. And the memory of the execution of Charles I in 1649 was also clear. Politics, religion, and rats had a new lease on life.

King Rat: Rodent Religion and Politics

In the Middle Ages, there was a saint for almost everything, and rats shared in the religious bounty. A seventh-century Flemish nun, St. Gertrude of Neville (629–59), is considered the patron saint of rats and mice, with whom she is often depicted—they represent the souls she was said to protect on their journey through Purgatory, or perhaps her protection of crops from their ravages. Another rat-focused religious personage was the German Bishop Hatto of Mayence in the tenth century, about whom no one had anything good to say. During a famine, instead of sharing his grain with starving peasants, he gathered them in a barn and burnt them alive, and, according to the account by Oliver Cromwell's teacher, the theologian Thomas Beard:

> But God that had regard and respect unto those poore wretches, tooke their cause into his hand, to quit this proud Prelate with just revenge for his outrage committed against them; sending towards him an army of rats and mice to lay siege against him with the engines of their teeth on all sides, which when this cursed wretch perceived, he removed into a tower that standeth in the midst of Rhine, not far from Bing, whither hee presumed this host of rats could not pursue him; but he was deceived: for they swum over Rhine thick and three-fold, and got into his tower with such strange fury, that in very short space they had consumed him to nothing; in memoriall whereof, this tower was ever after called the tower of rats. And this was the tragedy of that bloudy arch-butcher that compared poore Christian soules to brutish and base creatures, and therefore became himselfe a prey unto them.[36]

Calvinists during the English Civil War of the mid-seventeenth century did not hesitate to use lives of saints and prelates to vilify the Catholic Church. Bishop Hatto was an easy target, a sinful prelate justly destroyed by rats. (The cleric who sparked the Reformation, Martin Luther, described the apex of the hierarchy of the Catholic Church as "the Pope, the king of rats right at the top.") Beard also brought rodents into an attack on the Catholic mass, demanding, "Whether if a Rat eate the Host, he be thereby sanctified and, and made an holy Rat?[37] Calvinist authors condemned even the presumably sweet St. Gertrude, sneering that she was "worship by superstitious people, because (as they say) she preserves them from rats and mice."[38]

Linking theological enemies to rats was common to all sides in the religious wars of the seventeenth century. Sir John Denham (1615–69), an Anglican poet and royalist who lost his estates and position during the English Civil War, wrote a long satirical poem about English Calvinists trying to install their church in Ireland; it starred a rat named Rattamountain, "a Ratt of Fame ... He all the Rules and Tricks could show,/Both Arts of War, and Peace did know,/To cheat a Friend or spoil a Foe." Deciding to go to war,

Rattamountain rallies his forces by invoking the rodent victory over Bishop Hatto: "And did we not neer Mentz devour, / Their Prelate (Maugre [in spite of] all his power)." The rats will triumph, he promises, because they are so fertile their females will produce an endless supply of soldiers, and, moreover, the rats will ally with mice against the Irish cats. In their newly formed Presbyterian Parliament, the rats will become the peers and the mice the commons, and Rattamountain will be king. All the playhouses will close (as Cromwell did in England) and be replaced by conventicles.[39]

Denham's rats possess many of the negative qualities associated with rats and with people having rat-like characteristics: they are cunning, incredibly fecund, and upset the traditional balance of nature and society. Both sides of the English Civil War, the royalist Cavaliers and the parliamentary Roundheads, were driven by religious passions, producing atrocities with rat-like resonances. One parliamentary supporter charged that William Smith, Charles I's keeper of the prison in Oxford, threatened Roundhead prisoners that unless they signed a royalist loyalty oath, "he would make us to shit as small as a Rat."[40] Presumably, that was because their prison diet would produce a certain excretory result, but the threat also speaks to the view that enemies were a kind of rat. Once again, dehumanizing the other means verminizing them.

Oliver Cromwell even saw rodent identity in a supporter Edmund Harvey, who "being accused of fraudulent dealings . . . was discarded by Cromwell. . . . I never heard any that could speak of his honesty or courage, being as to the last a little inconsiderable ratt, and as to the other a factious Rumper, and one of his Majesties cruell Judges."[41] After switching sides several times, Harvey ultimately supported parliament against the king but was afterward convicted of malfeasance and stripped of the offices he had been granted by Cromwell. At the Restoration, he was denied indemnity and spent the rest of his life in prison.

If Charles I had known of Harvey's fate, he might have taken some satisfaction at this rat getting his just reward. In the *Eikon Basilike*, a spiritual autobiography attributed to the king and published in 1649 just ten days after his beheading, Charles supposedly wrote, "I see Vengeance pursues and overtakes (as the Mice and Rats are said to have done the Bishop in Germany) them that thought to have escaped and fortified themselves most impregnably against it both by their Multitude and Compliance. Whom the Laws cannot, God will punish, by their own Crimes and hands."[42] Thus awareness of Bishop Hatto and his rat-bitten finish reached even to the highest seat of power, at least until it became an ejection seat. The regicides are rats, rodents who have deserted their sovereign; they are, according to the spiritual auto-biography, "guilty of prodigious insolencies; when as before, they were counted as Friends and necessary Assistants."[43] Or perhaps they are rats, as is their proverbial custom, deserting a sinking ship or a burning house.

By 1660, the fire seemed to have been put out. The Restoration of Charles II opened a period of peace and seeming tolerance for religious differences. The king was a pragmatist who avoided open conflict on religious questions. But many of his subjects did not share his temperate attitude. Serenus Cressy, an English convert to Catholicism who later became a Benedictine monk, accused the Anglican clergyman Edward Stillingfleet of being like a rat who "foresees, or shrewdly suspects some danger to the Ship [of state]" and aban-dons it and "therefore provides for his own safety, by returning to the same Sects which incessantly plot against it," notably the radical Protestants re-sponsible for the execution of Charles I.[44] Not surprisingly, Stillingfleet an-swered by linking Cressy with the Jesuits—not popular in England at any time—who "like Rats have forsaken a sinking ship? It would be a great Joy to the whole Nation, to hear we were so well rid of them."[45]

Rats raised their snouts politically again after the Glorious Revolution, when the Catholic James II was booted out in favor of his daughter Mary and her husband, William of Orange, who became joint sovereigns of England. In a fable written in 1698 but credited to Aesop, "The Weesil, Rats, and Mice," a crafty and power-hungry weasel-king decides to augment his power by de-stroying the mice who might challenge him. He consults an "aged Rat"

> And ask'd him his advice,
> Whether a Project mayn't be try'd
> To eat up all the *Mice.*
> Ay, quoth the *Rat,* your Majesty
> May be well satisfy'd.
> *Mice* haters are of Monarchy,
> And Regal State deride.

Together, King Weasel, his weasel followers, and the rats devour all the mice. But it was a bad bargain for the rats, which the weasels now devour:

> Kings must have sumptuous Meat.
> The Rats now all do go to pot:
> Some Bak'd, some Boil'd, some Roasted;
> 'Tis hop'd they had not then forgot

How they the Mice accosted.
Thus some Men oft by Tyrant Power.
Their Kindred, Subject Slaves devour,
Do all the Villanies are done
To prop a beastly Tyrant Throne;
Tho' others Blood the Tyrant fill'd,
They must at length to's Fury yield;
Nought stops a Tyrant's Course but Decollation,
Or else a modern Abdication.[46]

So, tyrants get their due either by decollation—meaning decapitation—or abdication, after they seek to destroy their subjects. Such were the fates of Charles I and James II.

Rats continued to surface politically, but more as a metaphor for inept or corrupt government than for doctrinal controversy. This is particularly evident in the response to the new Hanoverian dynasty that began in 1714 with the accession of George I, the German-speaking Elector of Hanover. Unfortunately for him, his accession coincided with the spread of the brown or Norway rat, which rapidly displaced the black rat and came to be called by royal critics of the monarchy: the Hanover rat.

In his parody of a learned scientific treatise, *An Attempt towards a Natural History of the Hanover Rat*, published anonymously in 1744, Henry Fielding capitalized on the brown rat's appearance to attack the second Hanoverian king, George II (r. 1727–60).[47] His narrator begins by mentioning that this type of rat first appeared about thirty years before the treatise's publication—in 1714—the date of George I's accession. The narrator explains that the rat came from Germany and its favorite food is pumpernickel, and although it was "very lean," its voracious appetite soon makes it grow to the size of an elephant. In fact, the more it was fed, "the more greedy it grew," especially devouring the goods of "the middling rank" and country gentlemen.[48] In the complex English politics of the mid-eighteenth century, these were the groups that would generally oppose Robert Walpole, the first English Prime Minister, who dominated English politics between 1721 and 1742.

According to Fielding's satire, the Hanover rat hoards what it takes and "will not allow an English Rat, if it were starving, to touch one of their Hoards."[49] The Hanover rat is aided in its depredations by English accomplices or "Providers," and together they "immediately fell to undermining a large Building near the Thames, which was said to be raised upon such a solid

Foundation that nothing can hurt it; and has always before served as a safe Retreat for our English Rats when exposed to any extraordinary Danger."[50]

The large building is clearly the Houses of Parliament, where its English ministers now aid the Hanover rat in its voracious gulping of the resources of the country. It seems, our scientific gentleman argues, that England has lost all the ratcatchers who previously would have poisoned any rat "as have dared to peep out of its Hole, after it had taken any Thing from the Granary, Warehouse, or Shop; but now these Hanover Rats swarm to such a Degree, that we may truly say, the Farmer ploughs and sows, and the industrious Part of the Nation labour to feed Rats."[51]

The unpopularity of the Hanoverians and their English allies surfaced in Fielding's better-known *Tom Jones*, published in 1749 under his own name, in a speech from Squire Allworthy, Tom Jones's kindly benefactor:

> Pox! The world is coming to a fine pass indeed, if we are all fools except a
> parcel of roundheads and Hanover rats. Pox! I hope the times are a coming
> when we shall make fools of them, and every man shall enjoy his own. . . . I
> hope to zee it, sister, before the Hanover rats have eaten up all our corn, and
> left us nothing but turneps to feed upon.[52]

The Hanoverian monarchs also caused another eighteenth-century literary heavyweight to think of rats eating away at the resources of the kingdom. In "A Letter to Mr. Harding," published anonymously in the *Drapier's Letters* (1724), Jonathan Swift denounced the corruptly acquired grant of a patent to William Wood to mint copper coins to use in Ireland. "It is no loss of honour to submit to the lion," declared the letter, "but who, with the figure of a man, can think with patience of being devoured alive by a rat."[53]

In *Gulliver's Travels*, published two years later in 1726, the author's protagonist, Lemuel Gulliver, does indeed fight rats, presumably reducing him to the status of the rodent. In the novel, Gulliver journeys to the Kingdom of Brobdingnag, where he is captured by a farmer, made into a kind of toy by his daughter, Glumdalclitch, and put on display for the entertainment of the Brobdingnagians. While in their custody, he does battle with two rats:

> While I was under these circumstances, two rats crept up the curtains, and ran
> smelling backwards and forwards on the bed. One of them came up almost to
> my face, whereupon I rose in a fright, and drew out my hanger [his sword] to
> defend myself. These horrible animals had the boldness to attack me on both
> sides, and one of them held his fore-feet at my collar; but I had the good for-

tune to rip up his belly before he could do me any mischief. He fell down at my feet; and the other, seeing the fate of his comrade, made his escape, but not without one good wound on the back, which I gave him as he fled, and made the blood run trickling from him.[54]

The rats are the epitome of every awful thing that can happen to a man who has devolved to the level, or at least the size, of a rodent. Gulliver's similarity to a rat becomes even more evident after he has been taken to the royal palace. The queen commissions "her own cabinet-maker to contrive a box" to serve as the traveler's bedchamber. The box is "sixteen feet square, and twelve high, with sash-windows, a door, and two closets, like a London bed-chamber. The board, that made the ceiling, was to be lifted up and down by two hinges, to put in a bed ready furnished by her majesty's upholsterer."[55] The ceiling is fitted with a lock to prevent rats and mice from coming in or perhaps to prevent the prisoner from getting out, since it is has a lock that Glumdalclitch secures every night. In other words, it is essentially identical to the cages that ratcatchers used to display their captives and prove their skill at catching rats, like the example on page 192.

Swift himself felt like a rat due to his self-described exile in Ireland as dean of St. Patrick's Cathedral in Dublin. He wrote to his friend Henry St. John, Viscount Bolingbroke, "It is time for me to have done with the world . . . and not die here in a rage, like a poisoned rat in a hole."[56] One of the reasons Swift was fated to die in Ireland like a rat in a hole was his antipathy for the political leaders of his time. Like Fielding, he had no love for Robert Walpole, although he did support the Hanoverians for religious reasons. In *Gulliver's Travels*, the Brobdingnagian king informs Gulliver:

> you have made a most admirable panegyric upon your country; you have
> clearly proved, that ignorance, idleness, and vice, are the proper ingredients
> for qualifying a legislator; that laws are best explained, interpreted, and
> applied, by those whose interest and abilities lie in perverting, confounding,
> and eluding them. I observe among you some lines of an institution, which,
> in its original, might have been tolerable, but these half erased, and the rest
> wholly blurred and blotted by corruptions.[57]

And so the king finds "your natives to be the most pernicious race of little odious vermin that nature ever suffered to crawl upon the surface of the earth." Even an imaginary king in an imaginary place knows the power of calling people vermin.

A primitive late eighteenth-century rat cage, ca. 1780. Courtesy of M. Charpentier Antiques

Political rats became an even more popular device of British satirical discourse in the later eighteenth century. In an anonymous caricature (page 193), Lord Frederick North, the English Prime Minister from 1770 to 1782, is portrayed as a ratcatcher, holding a box that would have made Gulliver feel at home.

The text of the magazine explains that this Political Rat Catcher may seem to be imaginary but actually exists and "is a person of no small consequence in the state. Indeed, when it comes to rats, he "can gratify the most voracious, tame the wildest, and silence the most noisy," with a powder that contains "the elixir of office, or the essence of pension." He is wily enough to let the rats loose periodically and establishes his reputation by quelling the resulting tumult. If anyone wants to see this remarkable personage, "he may be almost certain of meeting him any day during the sitting of Parliament."[58]

Lord North was not the only politician to get the rat treatment. John Rob-

In Hoc Signo Vinces

The Political Rat Catcher, from *Oxford Magazine* 8 (1772): 225–26. Courtesy of the Trustees of the British Museum

inson was a member of parliament and secretary of the treasury under Lord North but did not follow him into the opposition in 1782 when the American Revolution drove Lord North from office. Robinson's followers were called "Robinson's Rats." The illustration on page 194 indicates him watching them contentedly while he catches them in his traps, labeled with the various enticements he used to gain their support: promises of advancement in rank for the naval officer, a seat in parliament for another, and a pension of £1,000 for a third supporter. A shredded Magna Carta is pinned up on the wall,

The Political Rat Catcher, or Jack Renegado's New Patent Traps, 1784. Library of Congress Prints and Photographs Division, http://loc.gov/pictures/resource/cph.3a31591/

while a drawing of William III is covered with a cobweb. The political rat-catcher is evidently happy to desert his previous party for his own political advancement—he is a rat seeking the patronage of the new prime minister, William Pitt the Younger. The poem underneath the picture reads:

> Thus when Renegado sees a Rat
> In the traps in the morning taken
> With pleasure he goes to Master Pit to pat
> And swears he will save his Bacon.[59]

Satirists, including Swift and these cartoonists, were well aware of the negative stereotypes connected to the rat-catching profession. Ratcatchers were the lowest of the low in the English social hierarchy. As early as the seventeenth century, the dramatist John Day had a character ironically describe his "good breeding": "My great Grandfather was a Rat-catcher, my Grandsire a Hangman, my father a Promoter, and my selfe an Informer."[60] Colloquially, a ratcatcher was someone who ratted on or destroyed another person for the benefit of himself or the authorities. Like hangmen, and those who informed on their neighbors, they were marginalized figures, sometimes equated with rogues and thieves.[61] As we saw earlier in the ballad "The Famous Ratketcher," that ratcatcher "can Colloque with any Rogue,/and Cant with any Gipsie." He also "Full often with a Negro,/The juice of Poppies drunke hee." The racist connotations attached to the figure of the ratcatcher are clear here; by associating with Black people and Gypsies, the Ratketcher is put on a level with those most despised in English society. The Ratketcher is also such a drinker: "it is doubtfull whether/He taught the Rats, or the Rats taught him/to be druncke as Rats together." This analogy should be read as the equivalent of the more recent "drunk as a skunk," but it also implies that the ratcatcher gets down and dirty with the rats, almost becoming one of them.[62]

The negative stereotype of ratcatchers continued into the nineteenth century, when the sport of rat baiting or ratting became popular after 1835, when baiting larger animals was prohibited by an act of parliament. It was the rodents' aggressiveness that led to rats becoming something like entertainers. We saw that flea circuses subjugated fleas but also portrayed them in heroic, or at least benign, ways. Rats, however, were used to satisfy the bloodlust of their audience, forced into rat pits and massacred by dogs, while viewers bet on how many the dogs could kill, and how quickly.

The most famous account of this bloodthirsty activity was by the journalist and social critic Henry Mayhew, the interviewer of Mr. Tiffin, the Bug Destroyer to Her Majesty. In *London Labour and the London Poor*, the proto-social scientist spent many pages describing ratcatchers, including a sewerman "who combed through the muck to find rats to sell for the rat pit." This man's house "is swarming with children," a veritable rat's nest, and the sewer man's "eyes have assumed a *peering* kind of look, that is quite rat-like in its furtiveness."[63]

But Mayhew also recognized that rat catching was becoming professionalized, mimicking the rising occupation of exterminator in the eighteenth century. Professional ratcatchers were eager to claim membership in the rising middle class, negating their former bad image. Mayhew interviewed at length the ratcatcher Jack Black, who described himself as "Rat Catcher to the Queen" and dressed accordingly.

Jack Black's sash, decorated with rats (cast from his wife's copper pots—she was not happy about that), reads "V-R," for Victoria Regina. He began his career as a child around 1800, catching rats around Regent's Park, then still meadows and fields, and three times he nearly died from rat bites. But by the

Jack Black, Rat Catcher to the Queen. From Henry Mayhew, *London Labour and the London Poor*, vol. 3 (London: W. Clowers and Sons, 1861), 11. Available from the Perseus Digital Library

time he started supplying rats for the rat pits in the 1830s, he had learned how to handle them: "I found," he told Mayhew, "I was quite the master of the rat, and could do pretty well what I liked with him." Realizing he could prosper from his trade, both by killing and selling rats, Black had his costume made and started selling rat poison from a cart, perfecting an act during which he began "the show by putting rats inside my shirt next to my buzzum [bosom], or in my coat and breeches pockets, or on my shoulder. . . . I used to handle the rats on every possible manner, letting 'em run up my arm, and stroking their backs, and playing with 'em."[64] These tamed rats, not realizing that doom is upon them, anticipate a happier fate as beloved pets.

Jack Black seems like a character out of Charles Dickens, one of those seeking to rise in social status. Dickens knew of Henry Mayhew's work, as did William Makepeace Thackeray, who has a character in *Vanity Fair* who learned about blood and breeding from rat killing: "O as for that," said Jim, "there's nothing like old blood; no, dammy, nothing like it. I'm none of your radicals. I know what it is to be a gentleman, dammy. See the chaps in a boat-race; look at the fellers in a fight; aye, look at a dawg killing rats,—which is it wins? the good blooded ones."[65]

Jack Black knew how to breed rat-killing dogs. He had one named Billy, whom he claimed had been the best rat killer in London and was the ancestor of all the city's really good rat-killing terriers. Indeed, he had sold one of Billy's progeny to the Austrian ambassador and another to the wife of a banker, presumably as pets.[66] But the audiences for Jack Black, his dogs, and his rats were usually members of the lower classes. He was part of the performing world, like the flea impresarios who staged spectacles for the amusement or edification of their fellows. When he dipped his hand into a basket of rats and brought them to climb around his body, it was "a feat which generally caused an 'oh!' of wonder to escape from the crowd."[67] As we have seen, Victorian London was the site of circuses, freak shows, and public exhibits of exotic animals and peoples, so Jack Black fit right into this particular atmosphere. He was a kind of monster who could perform superhuman feats, and he knew his rats as well as he knew himself.[68]

Jack Black does seem an amalgam of human and animal. He seems to embody Charles Fothergill's description of rats in *An Essay on the Philosophy, Study and Use of Natural History* (1813) Fothergill repeats all the characteristics of rats that colored earlier natural history and folklore. "The male rat has an insatiable thirst for the blood of his own offspring . . . a single male of more than ordinary powers, after having overcome and devoured all competitors with the exception of a few females, reigns the sole bloody and much-dreaded tyrant over a considerable territory." He notes that they are foreign invaders and almost impossible to kill.[69]

Likewise, Jack Black in his own estimation was "the master of the rat" who dominated the rodents like dominant male in a pack of rats. He was smart and cunning like a rat and enjoyed playing tricks on those unfamiliar with his talents. He was cannibalistic, at least when it came to rats, informing Mayhew that rats "were as moist as rabbits, and quite as nice."[70] Edible rats, for Jack Black at least, were no longer vermin, and eating them was a sign of civilization rather than savagery, although no Victorian cookbook contains rat recipes. Black also encouraged rat procreation, crossbreeding rats to produce "the finest collection of pied rats which has ever been knowed in the world. I had above eleven hundred of them—all wariegated rats, and of a different specie and colour, and all of them in the first instance bred from the Norwegian and the white rat, and afterwards crossed with other specie."[71] He sold these specialized rats as pets: his rats crossed the boundary between carnivorous animal and human play toy.

Like the species-melding rats he produced, Jack Black was a boundary crosser who attempted to rise above the class into which he had been born. His rat catching was so successful he was able to buy a tavern. His daughter served behind the bar, dressed as the "Ratketcher's daughter in velvet and lace, with a muslin skirt and her hair down her back, she looked very genteel, added the parent."[72]

Unfortunately, the tavern did not thrive, and Jack Black was forced back into the rat-catching trade. He caught rats for members of all classes and places in London because, as would be true throughout history, "Rats are everywhere about London, both in rich and poor places. I've ketched rats in 44 Portland-place, at a clergyman's, house there." He also aided a "medical gent" in Hampstead whose children had been bitten so badly by rats "their little night-gownds was kivered with blood, as if their throats had been cut." After Black destroyed the rats, "when I used to pass by that there house, the little dears when they see me used to call out to their mamma, 'O, here's Mr. Ratty, ma!'"[73]

Mr. Ratty, indeed. When Henry Mayhew recounted the life and exploits of Jack Black, he painted a picture of a society and its beliefs, what he called "the undiscovered country of the poor." Mayhew was attempting a description of all the various characters inhabiting the netherworld of London, "supplying information concerning a large body of persons, of whom the public has less knowledge than of the most distant tribes of the earth."[74] How much Mayhew shaped the stories he heard, and whether he was sympathetic to the people he was portraying, is open to debate.[75] He was certainly pleasantly surprised by Jack Black, who was "a very different man from what I had expected to meet, for there was an expression of kindliness in his countenance, a quality which does not exactly agree with one's preconceived notion of rat-catchers."[76]

The preconceived image of ratcatchers and the vermin they caught was usually negative, even if the craftiness rats displayed was admired. In the sixteenth and seventeenth centuries, rats were the familiars of witches who had given themselves to the devil and joined his realm, like the Leicestershire witches or the witches in *Macbeth*, or because they literally come from another land. As we saw in chapter 4, Scotland and Ireland were both places where "the other" lives. Likewise, when ballads and fables tell the stories of rats and people, they are located in Germany with poor Bishop Hatto or Rattamountain's kingdom in Ireland or the exotic lands where the Ratketcher wanders with gypsies and Africans, or Gulliver's Brobdingnag. When foreign brown rats invade England, they are brought to the country by the foreign Hanoverians, who in turn encouraged those most rat-like among their new subjects and undermined the traditional institutions of the state. The monarchs are aided and opposed by political ratcatchers, the epitomes of aliens in a society. Actual ratcatchers continued their trade into the nineteenth century, when they gratified the most bloodthirsty of London's population by supplying rats to the rat pits. In their way they were as alien as the foreign ratcatchers who proceeded them. Henry Mayhew interviewed one of the first proprietors of a rat pit in England, one Jimmy Shaw: "The poor people," said the sporting landlord, "who supply me with rats, are what you may call barn-door labouring poor, for they are the most ignorant people I ever come near. Really you would not believe people could live in such ignorance. Talk about Latin and Greek, sir, why English is Latin to them—in fact, I have a difficulty to understand them myself."[77]

From all directions, the very fabric of English society could be gnawed by rats—and by the rat-like.

CHAPTER NINE

The Two Cultures of Rats, 1800–2020

In Bram Stoker's 1897 novel *Dracula*, a madman whose tastes include drinking the blood of flies describes his encounter with Count Dracula:

> Then he began to whisper: "Rats, rats, rats! Hundreds, thousands, millions of them, and every one a life; and dogs to eat them, and cats too. All lives! all red blood, with years of life in it; and not merely buzzing flies!" . . . He beckoned me to the window. I got up and looked out, and He raised his hands, and seemed to call out without using any words. A dark mass spread over the grass, coming on like the shape of a flame of fire; and then He moved the mist to the right and left, and I could see that there were thousands of rats with their eyes blazing red—like His, only smaller. He held up his hand, and they all stopped; and I thought he seemed to be saying: "All these lives will I give you, ay, and many more and greater, through countless ages, if you will fall down and worship me!"[1]

The poor man does indeed worship him, and his fate is not pretty. In succumbing to the master of rats, he represents the many people who experienced rats firsthand in the millennia before the twentieth century. Rats were omnipresent in the fields and towns of Europe, and the efforts to control them crept into society, religion, politics, and literature. Like the vermin we have examined in the previous chapters of this book, rats were boundary crossers, outliers of the natural world who seemed to enjoy living with human beings, eating their crops, invading their homes, and biting their children. Next to domesticated animals, some of which were used to hunt and destroy them, rats were the mammals most familiar to people. When the inhabitants of farms and towns moved into the urban areas of Great Britain, the rats moved with them, exchanging hayricks for sewers and basements. Just as Count

Dracula and his minions moved from the Transylvanian countryside to London, country rats became city rats and prospered, sucking sustenance and wreaking havoc.

More than any other vermin, people see themselves in rats, which have been inescapable figures in the human landscape. Wherever there are people, there are rats feeding on human detritus but also, for at least the past two centuries, aiding scientific experimenters and pet shop owners. Smart, endlessly fecund, and adaptable, the rat is a key to understanding the fears and affections of the modern world.

We have seen that the projected images of bedbugs, fleas, and lice allowed conquerors and masters to denigrate, persecute, and even exterminate those who were felt to be "the other." Rats are also about those we hate, but they evoke more complex visions of ourselves and our capacity for good and evil. As pets, rats testify to our affectionate natures; as vermin, they encapsulate all the moral depravity of the individual and society.

The rat is bound up with civilization; it flourishes in civilized spaces—the home, the city, the laboratory—but can swiftly symbolize the fall of order: the overrun restaurant, the bombed city, the urban wastelands abandoned to the homeless and the mad, and the human mind turning on itself. But some people perceive rats as cute—at least the small white version—and make them into pets. Their intelligence makes them a prime candidate for literary tales, particularly those with an audience of children or young adults. And like fleas and lice, they can be used to puncture the presumption and power of supposedly mature adults.

Rats' looming presence reveals our obsessive curiosity about these creatures, whose actions often seem to mimic human behavior. Popular expressions testify to their impact on the human psyche. Centuries-old expressions claim that rats will desert a sinking ship and a burning house. James Cagney famously, although apocryphally, called another gangster "you dirty rat." Michael Cohen, a former lawyer of Donald Trump's, is called a rat by conservative commentators, someone who ratted on his boss and friend.[2] The *Washington Post* liberal columnist Dana Milbank labels Trump as "Rattus Potus," who "is highly aggressive with other rats (except for immediate family members), often for no apparent reason," and Mitch McConnell as "Rattus ginormous," a member of a "powerful, beady-eyed species."[3] Rats burrow into human consciousness, symbolizing all that is sly and disgusting in the human experience.

But rats, like humans, are also recognized as smart and social, attested by the experimenters who employ them. Curt P. Richter (1894–1988), a psycho-biologist and geneticist who developed a poison for killing rats during World War II—and was known as the Pied Piper of Baltimore—enthused, "We all learned to have a high regard for the wild rat—its aggressive fighting spirit, high intelligence, and alertness."[4] Domesticated rats may lose some of their fighting spirit, but scientists argue they can become cooperative and even altruistic, putting together their superior intelligence to help each other.[5]

It was their aggressiveness, however, that led to rats becoming something like entertainers. We saw that flea circuses subjugated fleas but portrayed them in heroic or at least benign ways. Rats, however, were used to satisfy the bloodlust of their audience when forced into rat pits and massacred by dogs, while viewers bet on how quickly and how many the dogs could kill.

Few rats survived the ordeal of the rat pit, but rat intelligence, like human intelligence, allowed the creatures to escape the snares which people set for them. Exterminators agree that it is impossible to totally exterminate all rats, particularly since rat fecundity outpaces rodenticidal poisons—to which they also develop immunity over time. Bobby Corrigan, one of the nation's foremost experts on rats and how to get rid of them, testifies, "Rats are very incredible, wildly intelligent mammals, and human beings keep going around trying to exterminate [them] as if it's the opposite."[6] Rats' legendary survival of nuclear tests in the Solomon Islands bolsters the belief—or the fear—that rats will inhabit the earth long after humans have destroyed themselves.

And while human beings are fragile, rats are not. At night, human beings are vulnerable in their sleep, while rats are nocturnal animals, biting babies and children in their beds. They are also the Jeffrey Dahmer of animals, cannibalizing their weaker members when the opportunity presents itself. They seem like the epitome of evil, and, as in previous centuries, they are often considered demonic.[7]

In Robert Browning's "The Pied Piper of Hamelin," the poet links Brunswick, where the infamous incidence happened, with Hanover; these are Hanover Rats that the Pied Piper lures into the Weser River. This ratcatcher, genderless and foreign, "With sharp blue eyes, each like a pin, / And light loose hair, yet swarthy skin / No tuft on cheek nor beard on chin, / . . . ; / There was no guessing his kith and Kin," agrees to kill the rats for a thousand guilders, which the mayor and the corporation agree to pay him. When the mayor defaults on the payment to "a wandering fellow / With a gipsy coat of red and yellow!" the Pied Piper pipes all of the children of the town to a portal through Koppelberg Hill. Strangely enough, since it's nowhere near Hamelin, the children end up in Transylvania and become the progenitors of alien, oddly dressed people, who we can reasonably assume are Gypsies, the group most "other" in Victorian England.

In Browning's poem, the Pied Piper told the mayor he rid the Nizam of India "of a monstrous brood of vampyre-bats" (l. 92). Who knows whether he would have been as successful dealing with Dracula, or whether he might even have joined him in attempting to bring bloodsucking rats to England— they both had intimate ties with Gypsies.[8]

Rats meant danger in early modern England, but by the end of the nineteenth century, the animals had been domesticated and made into pets. Rat-catchers were selling them to upper-class ladies to keep in their homes. Scientists were beginning to use them in in their labs. But the nefarious connotations of the rat continued and amplified the genocidal imaginations of the Nazis and the nativist and racist feelings of some paranoid people in recent times.

The Terror of Rats

In the twentieth century, rats were hated and feared, as the most famous scene in George Orwell's *1984* makes clear:

> The circle of the mask was large enough now to shut out the vision of anything else. The wire door was a couple of hand-spans from his face. The rats knew what was coming now. One of them was leaping up and down; the other, an old scaly grandfather of the sewers, stood up, with his pink hands against the bars, and fiercely snuffed the air. Winston could see the whiskers and the yellow teeth. Again the black panic took hold of him. He was blind, helpless, mindless.[9]

Poor Winston Smith. His attempt to resist the totalitarian state headed by Big Brother is thwarted by his fear of rats. Driven to the edge of madness by his interrogator threatening to unleash a cage of rats to eat his face, he rats on his lover and accomplice, Julia. The grandfather rat is one rat too many—the beast makes the man a slobbering example of inhumanity, a mindless victim of the animal he most fears.

Orwell wasn't the only twentieth-century giant to smell a rat at the core of human terror. Sigmund Freud, in *Notes upon a Case of Obsessional Neurosis* (1909), reported that his patient Rat Man declared that rats were entering into the anus of his fiancée and father and devouring them from the inside out. In his first published psychoanalytic report, Freud associated rats with sexuality, money, and death. The rat, he argues, in legend is "a chthonic [underworld] animal, . . . one might almost say; and it is used to represent the souls of the dead."[10]

Rats are always beneath us—in our toilets, under our subways, in the sub-basements of our homes and sewers. As such, they provide a perfect metaphor for the people we consider beneath us. By 1940, Nazis were using the image of the rat in their anti-Semitic propaganda efforts, just as they utilized the louse. In the film *The Eternal Jew*, images of swarming rats are juxtaposed with pictures of the Jews in the Lodz ghetto. In a propaganda poster in occupied Denmark, a Jew is depicted as a kind of rat and the caption says, "Rats: Destroy Them."

Today, immigrants and Muslims fill the role of rat. In 2005, after the American invasion of Iraq, the Italian journalist Oriana Fallaci wrote, the "sons of Islam breed like rats."[11] The sociologists Erin Steuter and Deborah Wills point out, "The images and language we use to discuss the war on terror have a powerful impact on the way we think about and treat other human beings. The metaphors we draw upon, are often profoundly racist, increasingly threaten our chances of building a safe society."[12]

In American media accounts, rats and people of color make joint appearances. A *New York Times* headline in 1964 announced, "Boy with Rat Bites Found Dead in a Tenement Flat in Brooklyn." The boy's mother is described as unwed and on welfare, details that conjure up images of race and neglect, particularly when the newspaper mentions three other incidents of rat bites in the African American neighborhoods of Coney Island (1959), Brooklyn (1962), and Harlem (1963).[13]

The connection persists. In 2017, in a focus group at the South Carolina Freedom Summit, the mother of Citizens United president David Bossie compared Hispanic immigrants to rats: during the focus group, led by GOP pollster Frank Luntz at the summit, she complained, "People are comin' in this country across the borders like rats and roaches in the wood pile."[14]

Curiously, rats also sniff the air in liberal diatribes. A 2014 post on the progressive website Daily Kos proclaimed, "Right out of college and motivated by student loans, a fresh generation of Dreamers enter The Great American Rat Race. Sponsored by Capitalism, the prize promised at the end of the rainbow, is that American Dream everybody is talking about."[15] The dreamers here are not immigrant children hoping for a better life as American citizens, but young people chasing after their own tails.

Perhaps these newly minted capitalist rats were brought up on stories about rodents that they read while cuddling their own pet rats. Children's literature and films are full of anthropomorphized rats displaying the range of human traits: devouring rats, thieving rats, smart rats, scientific rats, and civilized rats, rats who are social pariahs, and rats who are modeling anything humans are or might become. Rats may just be rats, but somehow they are also human, or at least what humans think about themselves.[16]

In Robert O'Brien's Newberry Medal–winning novel, *Mrs. Frisby and the Rats of NIMH* (1971), a scientifically enhanced rat tells his fellows:

> The real point is this: We don't know where to go because we don't know what we are. Do you want to go back to living in a sewer-pipe? And eating other people's garbage? Because that's what rats do. But the fact is, we aren't rats anymore. We're something Dr. Schultz has made. Dr. Schultz says our intelligence has increased more than one-thousand per cent. I suspect he underestimated; I think we're probably as intelligent as he is—maybe more. . . . Where does a group of civilized rats fit in?[17]

Nicodemus, leader of the fictional rats of NIMH, proposes an exodus from the soft life they are living in a vacated mansion, full of food and ease. Another rat dissents, arguing, "You've got this idea stuck in your head. We've got to start from nothing and work hard and build a rat civilization. I say, why start from nothing if you can start from everything? We've already *got* a civilization." To which Nicodemus replies, "No. We haven't. We're just living on the edge of somebody else's, like fleas on a dog's back."[18]

Science and Society

Where do rats fit in? In the mid-twentieth century, scientists began to equate rat and human behavior—rats were no longer just a metaphorical way to picture the other, but a genuine source of information about man. The scientist in O'Brien's book is modeled on an actual research psychologist and ethologist, John B. Calhoun (1917–95). From 1944 until his death, Calhoun experimented with rats and mice at Johns Hopkins and other universities.[19] He moved to the National Institute for Mental Health in 1955 and developed a highly influential theory about human overcrowding based on his study of rodents. In *Scientific American* in 1962, he discussed his experiments with rats confined to a small quarter acre of land (which smelled terrible) and allowed to eat and breed with abandon. To his surprise, instead of the colony reaching 5,000 members, it stabilized at 150 rats, populating a small and squalid section of the space that the scientist had once considered a "rat utopia." As they huddled together in what Calhoun called a "behavioral sink," aberrant and pathological behavior increased. Instead of producing healthy babies, many of the female rats "were unable to carry pregnancy to full term or to survive delivery of their litters if they did. An even greater number, after successfully giving birth, fell short in their maternal functions. Among the males the behavior disturbances ranged from sexual deviation to cannibalism and from frenetic overactivity to a pathological withdrawal."[20] Dominant males in this crowded territory were able to establish "a harem of females," while some of the subordinate males became homosexual or hypersexual. Others became "somnambulists" that looked "sleek" and "fat" but were totally disassociated from the rest of the rat community.[21]

In 1962, in the contemporary debate about the dangers of overpopulation and urban decay, Calhoun did not hesitate to draw parallels between his own studies and the current societal challenges:

> It is obvious that the behavioral repertory with which the Norway rat has emerged from the trials of evolution and domestication must break down under the social pressures generated by population density. In time, refinement of experimental procedures and of the interpretation of these studies may advance our understanding to the point where they may contribute to the making of value judgments about the analogous problems confronting the human species.[22]

It did not take long for others to see a connection between rat overcrowding and humans in inner cities, including Calhoun himself.[23] Many writers, social scientists, and even politicians utilized his ideas. Most uncritically accepted his notion of "the behavioral sink" because it confirmed their own suspicions about the influence of crowded conditions and moral decay in the urban environment.[24] The negative conclusions of Calhoun's work translated well into other mediums. The journalist Tom Wolfe, in an essay titled "O Rotten Gotham—Sliding Down into the Behavioral Sink," which appeared in *The Pump House Gang* (1968), described overcrowding in the language of rats and demographic disaster: "Overcrowding gets the adrenalin going, and the adrenalin gets them hyped up. And here they are, hyped up, turning bilious, nephritic, queer, autistic, sadistic, barren, batty, sloppy, hot-in-the-pants, chancred-on-the-flankers, leering, puling, numb." He continued by describing Grand Central Station: "The floor was filled with the poor white humans, running around, dodging, blinking their eyes, making a sound like a pen full of starlings or rats or something."[25]

Likewise, Lewis Mumford equated rat and human behavior in his classic *The City in History* (1968), "No small part of this ugly urban barbarization has been due to sheer physical congestion: a diagnosis now partly confirmed by scientific experiments with rats — For when they are placed in equally congested quarters, they exhibit the same symptoms of stress, alienation, hostility, sexual perversion, parental incompetence, and rabid violence we know find in Megapolis."[26]

John Calhoun did later suggest a non-apocalyptic potential to human development, owing to the creative and adaptive abilities of his homosexual and deviant rats; this aspect of his work was largely ignored. Calhoun thought, based on experimental evidence, that rats could develop a "culture"; that is, they could learn to live by the Golden Rule and help one another so that all could survive. They could be taught to have a conscience by responding to a bell. In this he believed that rat conditioning applies equally to people: "Human behavior is conditioned, too. . . . Once programmed, we have certain values and we behave accordingly. I see no difference between their programming and ours."[27]

Calhoun was not alone among midcentury scientists equating rat and human behavior. The Johns Hopkins psychobiologist and ratcatcher Curt P. Richter was a well-respected scientist who developed the concept of the biological clock and showed how the brain creates the body's circadian rhythms.[28] During World War II, he was employed by the National Research Council at the National Academy of Sciences, and then by the Office for Scientific Research and Development, to find a way to poison rats that the government feared would be used as agents in biological warfare. (As noted, the Japanese used bubonic plague delivered by fleas in porcelain bombs to kill thousands in Manchuria). Using a team of ratcatchers—which included Boy Scouts, engineers, air-raid wardens, lawyers, and anyone else who might take part in the "enjoyable" task of hunting rats—he field-tested rodenticides and finally developed alpha-naphthyl thiourea, or ANTU, which indeed did the job.[29] After the war, critics of his approach pointed out that the poisons were used in primarily African American and poor neighborhoods, where the residents were not warned of their possible effects on human health.

Richter's experience as a self-proclaimed "Reluctant Rat-Catcher" led him out of the laboratory and into the racial biases of his time. In "Rats, Man, and the Welfare State," he compared wild Norway rats to Australian aborigines whose lives "gave a picture of the life of man during the earliest stage of cultural evolution," while the characteristics of domesticated rats demonstrate how the welfare state has weakened humans, leading to disease and mental illness and heightened sexual activity. In the past, he argued, both wild rats and primitive humans were engaged in a Darwinian struggle for existence, but now "the necessities of life [for people] are assured just as fully as for the domesticated rat." The result, Richter propounded, was that the weak survived, weakening the population and potentially causing the downfall of modern civilized society. He cited historical precedents; in ancient Rome, "The weaker, less energetic individuals survived at an increasing rate, finally leaving a mass of individuals that no longer had the strength or will to fight for their country." In America, "there is a general increase in the number of people with defective physical and mental equipment as reflected by the high rate of rejections of men for service in the last war ... and the large numbers of breakdowns in concentration camps." Moreover, he argued, there may be a physiological component to this weakening of the human stock, which may be the result of changes in the adrenal glands, which "closely parallel those seen in the rat."[30]

Richter ended his account of rats and the welfare state by quoting eugenicists who warned about the evolutionary implications of allowing the weak to survive and whose views he clearly favors. "The knowledge derived from the study of the domestication of the rat," he argues, "may thus help us to face up to the important question: 'Where are we going?' and 'What is our destiny?'"[31] Especially if you think people—especially certain people—are a lot like rats.

But as one of his contemporary critics remarked, "While it is true that we spend only a fraction of the day in getting food, clothes, and shelter, we are not, like the laboratory animal, remaining all day in our comfortable beds exercising our overdeveloped gonads."[32]

Linking humans and rats had long expressed Americans' social fears. In the early twentieth century, immigrants from Eastern and southern Europe merited rat portraiture. The Library of Congress explains that the cartoon on page 209 "shows Uncle Sam as the 'Pied Piper' playing a pipe labeled 'Lax Immigration Laws' and leading a horde of rats labeled 'Jail Bird, Murderer, Thief, Criminal, Crook, Kidnapper, Incendiary, Assassin, Convict, Bandit, Fire Brand, White Slaver, [and] Degenerate,' and some carry signs that read 'Black Hand' showing a black handprint. In the background, rulers from 'France, Russia, Germany, Italy, Hungary/Austria, Turkey, [and] Greece', along with citizens of these countries, are cheering the fleeing rats."[33]

S. D. Ehrhart, *The Fool Pied Piper*, photomechanical print, 1903. Library of Congress Prints and Photographs Division, http://hdl.loc.gov/loc.pnp/ppmsca.26380

The Battle against Rats: War, Politics, and Race

The notion of rats swarming—symbolically—into a country and defiling it has a long and inglorious history that continues into the present.[34] The German silent film classic *Nosferatu* (1922) captures the notion that contagion—in the form of the vampire Nosferatu, an avatar of Dracula, and the rats living in his coffin—is invading from the East and bringing death to the West. It was produced in a period when there was widespread Jewish immigration from Russia and Eastern Europe after World War I and the Russian Revolution, often to the discomfort of westerners. The film prefigures the anti-Semitic propaganda used later by the Nazis, and the figure of Nosferatu, played by Max Schreck, has the large nose and ears often used in depictions of Jews.[35]

The Nazi film *The Eternal Jew* (1940) explained the connection between Jews and rats: "Where rats turn up, they spread diseases and carry extermination into the land. They are cunning, cowardly and cruel, they travel in large packs, exactly the way the Jews infect the races of the world."[36]

In the pages of *Der Stürmer*, Julius Streicher elaborated this connection:

The Jews are a people of bastards, afflicted with all diseases. They are a people of criminals and outcasts. They are the carriers of disease and vermin among men . . . A rotten apple cannot be assimilated by a basketful of healthy apples. Mice and rats cannot be acknowledged as useful pets and live within the community. . . . Bacteria, vermin and pests cannot be tolerated . . . for reasons of cleanliness and hygiene we must render them harmless by killing them off. . . . Why should we repress our feeling for cleanliness and hygiene when it comes to the Jew?[37]

World War II propaganda by the allies also identified the enemy as rodents. Wartime posters particularly portrayed the Japanese as a kind of rat. Some posters suggest that the Japanese may rape and kill white women, like this rat-whiskered Japanese soldier.

This Is the Enemy, poster from the US War Department. US National Archives, https://commons.wikimedia.org/wiki/File:US_propaganda_Japanese_enemy.jpg

The link among rats, warfare, and race long outlasted the war. In the 1960s, two *New York Daily News* headlines declared, "This Is It! We Pass Ammo to Troops of the Anti-Rat War" and "War Is On! Rat-Battlers Open Attack in E. Harlem!" which once again emphasize the racial underpinning of rodent attacks.

As American cities deteriorated over the twentieth century, the connection between their growing nonwhite population and rats deepened in the public mind. Some feared that this defective population might expand and corrupt healthy humans and monopolize government resources. Such attitudes influenced the work of Calhoun and Richter. Like rats, the inner city population was seem as overcrowded and oversexed, with pathological behavior that could endanger the rest of society.

Ironically, the novel *Native Son*, by the African American writer Richard Wright, most vividly portrayed the perceived connection between race and rats. The novel opens in the one-room apartment of Bigger Thomas's family in a slum in Chicago where a huge rat is threatening the family. The mother screams at her daughter, "Don't let that thing *bite* you" as she orders her son to kill the rat, which attacks him unsuccessfully. "The rat's belly pulsed with fear. Bigger advanced a step and the rat emitted a long thin song of defiance, its beady eyes glittering, its tiny forefeet pawing the air restlessly." Ultimately, Bigger kills the rat with an iron skillet, but his mother remains incensed at her son. She cries, "We wouldn't have to live in this garbage dump if you had any manhood in you."[38]

Wright uses the rat to express his characters' lives. They live in terrible, crowded conditions that allow giant rats to breed; Bigger and his brother speak in "tones of awed admiration" for the rat: "Gee, he's a big bastard." "That sonofabitch could cut your throat." "He's over a foot long." "How the hell do they get so big?" "Eating garbage and anything else they can get." In a sense, Bigger and the big rat are the same: aggressively fighting over a tiny bit of space and terrorizing others. Bigger takes the dead, "dangling rat, swinging it to and fro like a pendulum, enjoying his sister's fear." He is now the rat and he will kill, but in doing so he achieves a kind of power within himself. Although he will be executed, he tells his lawyer, "What I killed for must've been good! When a man kills, it's for something. . . . I didn't know I was alive in this world until I felt things hard enough to kill for 'em."[39] Earlier in the novel, the lawyer had warned that Bigger and those like him, both white and Black, are the product of poverty and oppression that "form the quicksands upon which the foundations of our civilization rest. Who knows when some slight shock, disturbing the delicate balance between social order and thirsty aspiration, shall send the skyscrapers in our cities toppling?"[40] Wright's message is that society must change or civilization is doomed—the foundations of our tallest buildings may be gnawed and undermined from the subbasements inhabited by humans and rats.

Front cover of George H. Smith's *The Coming of the Rats* (Pike Books, 1961). Illustration by Albert A. Neutzel, used courtesy of Charles Nuetzel

So it is understandable when the terror about rats erupts into popular culture. In a 1954 Marvel comic book, *The Chamber of Chills*, rats swarming in a sewer menace a helpless woman, defended by a courageous man. The cover screams, "One of the Most Spell-Binding Shockers of All Time! Death in a Long Dark Tunnel!"[41] Capitalizing on the fear associated with nature out of control, perhaps because it has been corrupted with by an atom bomb, this comic feeds into the same paranoia as *Godzilla* (about a giant dinosaur) and *Them* (about giant ants), both released in 1954, and another giant bug movie,

Tarantula! (1955). An illustration from a 1961 paperback explicitly links the terror of nuclear annihilation with rats. One of the rats is attempting to remove what's left of the woman's clothing, while the others feed on her body.

Even in a movie as benign as the 1954 Disney film *Lady and the Tramp*, the middle-class baby in the home of Jim Dear and Darling is almost attacked by a hideous rat and is saved by Tramp the dog, who obviously knows his way around lower-class threats. These rats are savage invaders that cross the border between wild and civilized spaces.

Rats carried the racist and classist undertones of popular culture into nonfiction as well as literature. In 1960, the journalist Joseph Mitchell published an essay, "The Rats on the Waterfront," explaining, "The rats of New York are quicker-witted than those on farms, and they can outthink any man who has not made a study of their habits. . . . They steal along as quietly as spooks in the shadows close to the building line, or in the gutters, peering this way and that, sniffing, quivering, and conscious every moment of all that is going on around them."[42]

In the 1960s, "spook" could mean what is does now—a ghost or apparition—but it was also, in the definition provided by the *Oxford English Dictionary*, "A derogatory term for a black person."[43] Even if the North Carolinian Mitchell wasn't consciously conflating rats and African Americans, at least some of his readers would catch the connection. Mitchell's rats lurk in the shadows and gutters—they are transgressive invaders of human spaces. But the people living in those spaces—for example, between 118th Street and Park and Eighth Avenues in Harlem—could use their rodent cohabitants to their own advantage. Robert Sullivan, in his study of rats in New York, recounts the story of Jesse Gray, a community activist in Harlem who, starting in the early sixties, organized rent strikes to protest the living conditions in the tenements and urged residents to "Bring a rat to court" and another to City Hall. For Gray and his fellow city dwellers, rats became allies in the battle against slum landlords. He proclaimed, "The tenants are like rats now. . . . Rats feel their power, and they come out in broad daylight and just sit there. Once the tenants feel their power, they stop running, they're not afraid anymore."[44]

In a rent strike, the solidarity of the group turns the indifference of the powers-that-be onto itself. The oppressed weaponize the essential power of the rat to evoke fear and disgust. But perhaps it's a victory without claws. In a recent study of rat populations in Baltimore, a DNA researcher found clusters of rats near the East Baltimore neighborhood where Johns Hopkins University is located and Calhoun and Richter worked.[45] This neighborhood has been targeted for redevelopment after decades of decay, but no one told the rats, who were disinclined to move from their homes—rats, like some humans, do not like change. One community activist, Glenn Ross, sees the emergence of a ghetto in East Baltimore as a long-term plan of city officials and Johns Hopkins, which wanted to buy up and demolish abandoned properties and then replace the African American population in order to create an "urban Renaissance." Reversing the association of rats and African Americans, Ross says he became an activist because he saw a rat, but "now here it is a few years later and I'm dealing with the two-legged rats."[46]

Another Baltimore writer also encountered the city's rat problem. The journalist Karen Houppert, who had recently relocated to "Charm City" (as Baltimore calls itself) and was teaching at Johns Hopkins, observed a dead rat festering on her street, to the general apathy of her neighbors. Trying to get a handle on the rat situation, she discovered that the city only faced up to it when trying to entice corporations and yuppies to move there. "If we are to continue to grow as a city," a city council resolution reads, "attract new residents, and to encourage citizens to invest where they live, it is imperative that we maintain our vigilance in controlling the rat population." She sums up her conclusions humorously, recognizing the rat role in government action:

> I attribute this to a population bulge—rat, not human. Here is my theory: Rat colonies get crowded, and territorial squabbles break out as one rat nation-state covets its neighbor's turf, until there is a world war. Fast forward a few months, and the rat warriors come limping home to procreate like mad. Enter the boomer rats. Where did all these unruly rats come from and what should be done about them, city officials suddenly demand—approximately every eight years.[47]

World war between rat nation-states—it seems like a job for a State Department undersecretary for vermin, or at least its conflict-resolution desk: In DC, "in some of the city's most densely populated areas, where trendy bars and restaurants have proliferated, complaints have increased at a staggering rate." Between 2013 and 2017, there was a 449 percent increase from Columbia Heights and a 430 percent increase from Capitol Hill. Mark Eckerwiler, another neighborhood activist, proclaims it "an eternal war."[48]

Other cities see similarly disconcerting numbers. Bobby Corrigan, the rat expert, thinks that global warming has spurred the enormous rise in rat sightings, which are always reported in battlefield metaphors. Robert Sullivan describes rats and man as engaged in "an unending and brutish war," echoing the rhetoric of the seventeenth-century philosopher Thomas Hobbes, who believed that human beings in the state of nature were engaged in a war of all against all, and human life was "solitary, poor, nasty, brutish, and short"—and possibly rat-infested.[49]

Hobbes believed that the only way to end this warfare in the state of nature was to create the all-powerful state. Such nations, however, in their relations with each other, duplicate the war of all against all. The image of the verminous and clever alien rat could easily be transformed into the alien other, an agent bent on destroying your society. During the McCarthy era in the United States, the Communist-hunting senator inquired, "Why worry about being fair when you are shooting rats?"[50] The vice president and future president, Richard Nixon, agreed. In a 1954 speech, he declaimed,

> Now I imagine that some of you listening will say: "Well, why all this hullaba-loo about being fair when you're dealing with a bunch of traitors?" As a matter of fact, I've heard people say they're a bunch or rats. What we ought to do is go out and shoot 'em. Well, I'll agree that they're a bunch of rats, but just remember this, when you go out to shoot rats, you have to shoot straight, because when you shoot wildly it not only means that the rat may get away easily, you make it easier on the rat.[51]

This convoluted statement may not clarify Nixon's feelings about McCarthy, but it does elucidate his feeling about rats—and Communists. Like rats, Communists should be shot (presumably metaphorically) carefully, because otherwise they will get away and continue their nefarious activities. In ideology as with race, the message remains is clear: kill the other that supposedly threatens your way of life.

Even Canada, the most courteous of nations, has viewed the feared outsider as vermin. In Alberta, rats came to play a role in provincial mythology. C. H. Douglas, the leader of the Social Credit Party, which dominated provincial government in the 1930s, reprinted *The Protocols of the Elders of Zion*, a fabricated document that described a plan by Jews to take over the world, and he referred to Jews as "parasites," echoing the fascist cry that they were vermin. By the 1940s, a party document proclaimed, "The events of the past year [1946] provide further evidence of a rapidly developing and pre-conceived plan for world domination, whose bold outlines are now so plain that many who formerly doubted are now thoroughly alarmed at the pattern of things to come."[52] Like rats, Jews creep in and threaten to take over.

In the 1950s, the Albertan political culture embraced a policy to keep

Norwegian brown rats out of the province. The discovery of oil and the corresponding growth of Calgary and Edmonton threatened the province's traditional agricultural culture. The vulnerability of Alberta to attack was projected onto the invasion by rodents. In posters produced by the Albertan Department of Agriculture, notes the art historian Lianne McTavish, rats represented more than simple rodent control: they carried reinforced the assumption among Albertans that they were particularly clean and free of the diseases that plagued other Canadians.[53] Indeed, one of the posters depicts the rat with teeth bared, evoking all the power of the association of vermin with the alien; it proclaims: "You Can't Ignore the Rats, HE'S A MENACE TO HEALTH: HE CARRIES GERMS – HOME: HE DESTROYS PROPERTY – INDUSTRY: HE CARRIES WASTE. KILL HIM! Let's Keep Alberta Rat-Free."[54]

The campaign appeared almost simultaneously with the popular culture illustrations that depicted ravenous rats and the movies portraying fearsome nuclear-altered animals. Rats and paranoia go hand in hand.

Rats threaten to destroy the homes, health, businesses, and integrity of Albertans. The literal threat is merged into the metaphoric message of purity against transgression, a defense of the virtue of the Albertans. Weakness on rats equates to treason, as civilization teeters before the foreign invader. The campaign commands both the collectivity and the individual to guard against the threat poised at the border—the health and freedom of all are at risk.[55]

Another Albertan poster, published in 1975, evoked the rhetoric and racism of the late nineteenth century. The text on the poster shouts, "The Only Good Rat is a Dead Rat."[56] In 1888, during a campaign against the Plains Indians, Gen. Philip Sheridan supposedly had proclaimed, "The only good Indian is a dead Indian," a sentiment later echoed by Theodore Roosevelt. Any resident of the largely white population of Alberta would understand the link between rats and First Nation peoples.

Rats are also linked with prisoners. Over decades, studies by neuroscientists demonstrate that prisoners in solitary confinement—the opposite of overcrowding—suffer brain and mental deterioration. University of Michigan neuroscientist Huda Akil argues that solitary actually can "literally shrivel areas of the brain, including the hippocampus, the region of the brain involved in memory, spatial orientation, and control of emotions."[57] Michael Zigmond of the University of Pittsburgh showed that the brains of rats kept in tiny, isolated cells become stressed, aggressive, and even unable to recognize other rats. Their brains begin to suffer also, and they develop disease more easily.[58] Humans confined in small, isolated cells, the scientist argues, have similar

Target: Bums, Chicago poster, April 20, 2015. Courtesy of Alisa Hauser, https://www
.dnainfo.com/chicago/

neurological symptoms, a kind of posttraumatic stress disorder that includes
detachment and aggression.

The fears of the time shape scientific research and the lessons we draw
from rats.[59] In the mid-twentieth century, the fear of overcrowding and slum
populations resulted in studies linking pathological behavior and race. In the
inwardly focused twenty-first century, a recent study of socially isolated rats
concludes that social isolation may lead to schizophrenia.[60]

Among the most visible populations associated with schizophrenia are
the homeless, released from mental institutions to wander aimlessly through
the streets of cities and towns. They are the rats of postmodern times, eating garbage and living under bridges and complicating the gentrification of
cities. A recent *San Francisco Chronicle* headline proclaims, "Poop, Needles,
Rats, Homeless Camp Pushes SF Neighborhood to the Edge."[61] In Chicago,
signs mimicking those used in a campaign against rats urged people to "Target: Bums," clearly picturing the homeless population as a kind of rat infestation. Both the rat and the homeless person wear the same snarl and the posters demand the same fate for them: "This area has been surveyed and baited
for the homeless (or indigent vermin) where needed as of 4/20/15."

London has seen a similar equivalence between the homeless and rats,
although in the bleak English vision, the homeless are eating rats rather than
being identified with them. The *Guardian* newspaper reported that Polish

workers, unable to get benefits, were barbecuing rats and drinking hand sanitizer.[62] They would have quite a feast, according to some accounts, because some of the rats were as big as two feet. The city responded by poisoning rats, not the immigrants, but the government has also been seeking another solution: sterilizing the rats, a project addressed by scientists at Edinburgh University.[63] Given rats' proclivity for procreation, it's open to question whether they would find this a more humane solution than rodenticides.

Paris also suffers from an abundance of rats, with many tourist attractions, including the Eiffel Tower, being inundated with the rodents. With the universal certainty that vermin are brought by outsiders, some Parisians blame the European Union and its requirement that rats be trapped rather than poisoned; others argue the rat increase is a result of global warming, causing the rising levels of the Seine.[64] In 2016, the city began a poisoning campaign to get rid of the enemy. In response, rat lovers sent a petition signed by twenty-six thousand people to the city government protesting the policy and urging the government to use rodent birth control instead. A French writer insisted, *C'est un animal de compagnie d'avenir, adapté aux petits logements.* "It's a pet of the future, adapted to small apartments."[65]

One rat activist, Claudine Duperret, leads a group called "Rat-Prochement," which takes in homeless rats, a French twist on the link between rats and homelessness. Some of her rats, she claims, are refugees from science labs. Others might be abandoned pets, initially owned by fans of the Pixar film *Ratatouille.*[66] The movie is definitely pro-rat—its protagonist is Remy, a rat with a fine sense of smell who longs to be a chef. His father accuses him of "thinking like a human," unlike rats, who "look after our own kind" and don't abandon their homes. By the end of the movie, rats and some humans have opened a restaurant together, hidden from the sight of the Parisian rat-catcher.

The sentiment that government officials and their capitalist allies oppress the vulnerable and weak, symbolized in art and literature as rodents, animates the work of the mysterious graffiti artist known as Banksy. In this mural, a mouse or rat, adorned with a Disneyfied Minnie Mouse bow, crouches before a wall inscribed with the legend "Mai 1968" (the 8 reclines on the rat's head), the year when Paris exploded in antigovernment protests. Banksy often portrays the oppressed as rats or mice, and this rat is rising in rebellion against the corporate symbol on her head, displaying the source of her oppression. Banksy has explained his use of rats: "If you feel dirty, insignificant or unloved, then rats are a good role model. They exist without permission,

Banksy, *Fifty Years since the Uprising in Paris*, 2018. Courtesy of Banksy

they have no respect for the hierarchy of society, and they have sex 50 times a day."[67]

Literary Rats and Pet Rats

Even literature supposedly aimed at children utilizes the metaphoric power of rodents, both mice and rats. Beatrix Potter was particularly fond of rats. She dedicated her story *The Tale of Samuel Whiskers or The Roly Poly Pudding* to her recently deceased pet rat:

IN REMEMBRANCE OF

"SAMMY",

THE INTELLIGENT PINK-EYED REPRESENTATIVE

OF A PERSECUTED (BUT IRREPRESSIBLE) RACE

AN AFFECTIONATE LITTLE FRIEND

AND MOST ACCCOMPLISHED THIEF[68]

In another children's story, Robert C. O'Brien's *Mrs. Frisby and the Rats of NIMH*, Mrs. Frisby is a mouse mother who turns to superintelligent rats to help her heal her sick son and to aid her in finding a home away from the danger of humans. These rats have escaped from Dr. Schultz's lab, where they had been contained in boxes stacked one on top of another, separated, tagged, experimented on, and injected with some chemical that makes them extremely smart. The procedure echoes the experiments of Dr. Josef Mengele at Auschwitz-Birkenau, whether the author of *Mrs. Frisby* consciously intended it or not. At the end of the book, Dr. Schultz plans to kill his escaped prisoners with cyanide: "They're coming with an extermination truck—cyanide gas, I think," explains the farmer on whose land they have been living.[69] The cyanide-based Zyklon B was the gas used by the Nazis to commit genocide. Rats and Jews coincide again. Happily, O'Brien's rats escape and, after a series of adventures, set out for a far-off valley where they can live without stealing, the aim of a complex plan of stockpiling goods and exploration of other areas.

The desire to live without stealing is also a motivating factor for the rats in Terry Pratchett's Carnegie Medal–winning novel, *The Amazing Maurice and His Educated Rodents* (2001). These rats have become brilliant and self-aware by inadvertently feeding on the trash behind the Unseen University, a school for educating wizards in magic. They are directed by a cat named Maurice who has gone through a similar transformation by eating an educated rat. He uses the rats in a Pied Piper–like scam to fleece townspeople who think they are being overrun by the rodents. The rats do not like to do this, explains a female rat named Peaches, because they have learned from a small, white, and deep-thinking rat named Dangerous Beans that it is "not morally right."[70] Dangerous Beans and his fellow educated rats had believed that humans and rats could live together because they thought that a children's book, *Mrs. Bunnsy Has an Adventure*, was a true account of the ways humans and rats interact. Clearly, this is a reference to Beatrix Potter and all other authors of children's literature who paint a rosy picture of the relationships between species. Maurice the cat points out that the true story of rat–human relations is a state of constant war, particularly because humans, unlike rats, like to make war. Banksy and Thomas Hobbes would approve of this declaration.

Almost every human assumption about rats—and conduct toward rats—is referenced or questioned in Pratchett's story. Breeding, stealing, intelligence, and sociability among the rat population are used to emphasize the relative nobility of rats compared to humans. But it seems like humans and rats might be able to live together in peace after mutually aiding each other. And that would be a nice way for the book to end, writes the author: "If it was a story, and not real life, then humans and rats would have shaken hands and gone on to a bright new future. But since it was real life, there had to be a contract."[71] The only way to end the war of all against all, man against rat, in this story is to create rules to govern their relations so they could then live happily ever after—or at least have their instincts constrained by law.

There is no such happy ending for a rat in J. K. Rowling's *Harry Potter and the Prisoner of Azkaban* (1999). The wizard who betrayed Harry's parents to Lord Voldemort has been hiding out as a pet rat named Scabbers belonging to Harry's friend, Ron Weasley. Scabbers is unmasked but retains his rat-like appearance in both the book and film (the transformation from rat to man is striking in the movie): "His skin looked grubby, almost like Scabbers' fur, and something of the rat lingered around his pointed nose and his very small, watery eyes." The fact that he was undiscovered in his rat form "is not much to boast about," says Sirius Black, the character who reveals the truth that Scabbers is actually a former friend of Lily and James Potter. Black tells Harry, "This piece of vermin is the reason you have no parents. . . . This cringing bit of filth would have seen you die too, without turning a hair. You heard him. His own stinking skin meant more to him than your whole family."[72]

Rowling's description of the rat-like human taps into all of the negative connotations of the rat fink, the stool pigeon who betrays his friends for his own advantage. He is filthy in body and soul. At first, Ron and Harry and their friend Hermione can't believe this accusation is true. He is a pet rat, and pets don't turn out to be traitors.

The rats so enjoyed by rat enthusiasts like Ron Weasley are referred to as "fancy rats" and may be descended from unusual rats saved from the rat pits by Jack Black, the "Rat Catcher to the Queen," and other rat hunters. It became fashionable to keep such animals as pets early in the twentieth century, and there are now fancy rat societies in England, America, and worldwide whose members meet to show their rats and discuss the particular traits of their favorite animals.

Many children and adults who have pet rats praise them and love them for many reasons. Lauri Serafin, a nurse in Seattle, lists ten different positive qualities pet rats possess, including the following: "Rat individuals have different personalities. Some are shy, mischievous, adventurous, curious, assertive or affectionate. . . . Rats form strong bonds in their social group and with caretakers." She warns, "It is cruel to have a solitary rat unless you can spend very large amounts of time with it. Rats do want to spend time with you and will flex their schedule to match yours."[73] The animal rights organization PETA agrees, listing the top ten reasons to cuddle a rat, including "They're sweeter than your grandma" and "They're cleaner than your college roommate."[74]

Beatrix Potter would agree. Rats will continue to inhabit the pages of literature, particularly children's literature. Little white rats share with their readers the capacity to be viewed as cute animals to be cuddled and not killed. In today's multicultural world, even their brown brothers inspire children. In one recent book, a traumatized ginger-colored rat has to wait to be adopted, unlike her white box mates who call her "Dirty, and stinky." But eventually she is chosen, and two nice white rats in her new home give "Abby time to adjust and learn to trust."[75] In David Covell's *Rat and Roach: Friends to the End*, a rat and roach, living together under Avenue A, replicate Neil Simon's odd couple in their personal habits; Rat calls Roach "Toothpick, Crabby Head, Flea!" and Roach calls Rat "Hair Ball, Tuna Breath, Mouse."[76] Apparently the worst insult in their world is to accuse each other of being another kind of vermin. By the end of the book, they learn to live together and to be the best of friends. The illustrations in this picture book, according to *Publisher's Weekly*, are reminiscent of Banksy's rats.

Morality is alive and well in literature aimed at children, but there is a more ambivalent message in stories meant for older readers. In T. Coraghessan Boyle's short story "Thirteen Hundred Rats," a man comes to love a white rat he has bought to feed his snake—and relieve his loneliness—and then buys thousands more to keep it company. They eventually eat his body, dead from pneumonia in his apartment. A friend asks in amazement,

> there must have been some deep flaw in him that none of us recognized—he'd chosen a snake for a pet, for God's sake, and that low animal had somehow morphed into this horde of creatures that could only be described as pests, as vermin, as enemies of mankind that should be exterminated, not nurtured. And that was another thing that neither my wife nor I could understand: how could he allow even a single one of them to come near him, to fall under the caress of his hand, to sleep with him, eat with him, breathe the same air?[77]

We are back in an Orwellian landscape, a world of nightmares where rats devour humans. Regardless of the rat's help in laboratory experiments or its place as a human pet, people are most often terrified of rats, which in their very identity seem to threaten the sanctity and sanity of human civilization. Often this threat is reified in attitudes toward those who are associated with rats or considered a kind of rat themselves, whether Jews, or the Japanese, inner city African Americans, immigrants, or people experiencing homelessness. Even science, in its search for truth, reflects social norms and the racism of its time, whether the fear of overpopulation by the racial other or by the isolating ennui of modern times. Rats are the sharpest reflection of our fears—and of ourselves.

The Power of Vermin

Democrats, Donald Trump tweeted judiciously, "want illegal immigrants, no matter how bad they may be, to pour into and infest our Country." Heading in the other direction, he demanded that four recently elected Democratic congresswomen of color should "go back to the totally broken and crime-infested places from which they came."[1] Nobody missed Trump's careful choice of verb. "Infested," mused Chris Wallace of Fox News. "It sounds like vermin."[2]

Indeed. "Infest" is the ultimate vermin verb. As David Graham of *The Atlantic* writes, "It drives full-throttle toward the dehumanization of immigrants, setting aside legality in favor of a division between a human us and a less-human them. What are infestations? They are takeovers by vermin, rodents, insects. The word is almost exclusively used in this context. What does one do with an infestation? Why, one exterminates it, of course."[3] When politicians use the term "infested"—and not that many of them would—it means more than simply "filling a space." It is vermin who infest, and vermin who must be removed.

What, in language or too frequently in action, transmutes human beings into vermin? The poisonous use of the term has a long and painful history. Vermin cross boundaries to invade our bodies, our homes, our countries. The term "vermin"—both insect and rodent—is wielded as a lethal metaphor against those we feel threaten humanity and civilization. Those described as vermin are not only dirty but also explicitly evil—their very existence threatens the rightful functioning of society. The list of the verminous has been long: they are immigrants, Jews, beggars, Scots, Irish, slaves, Africans, Native Americans, unruly women, Communists, and anyone else viewed as threatening the established order.

Vermin are disgusting and depraved—and their existence is uniquely useful for those seeking to make their own righteousness and superiority universal standards. In the Middle Ages, vermin were accepted as a sign of God's power, an attitude that persisted among some naturalists into the nineteenth century. But from the late seventeenth century onward, vermin—and the verminous—were increasingly seen as disgusting, as theological beliefs were undermined by class and racial attitudes.[4] The modern world, with its belief that civilized meant clean, was born.

During the recent Balkan wars, a Croatian ambassador wrote approvingly that a Serbian gunner was taking aim "at those scurrying ants, vermin to be cleansed off the face of Europe."[5] An Israeli rabbi urged about Palestinians, "What we should do is go into those vermin pits and take out the terrorists and murderers. Vermin pits, yes, I said, vermin, animals."[6] After the assassination of the Isis leader Abu Bakr al-Baghdadi, the *New York Post* commended the American operation for "hunting down the worst vermin, like Baghdadi."[7]

The label of vermin makes dehumanization possible, and dehumanization is the step next to violence. Beneath the labeling of the other as vermin is clearly fear—the anxiety that the suppressed will rise up and overthrow the righteous. Fear of insects and rodents turns into fear of the people they symbolize. Just as rats can penetrate gentrified neighborhoods, or bedbugs, lice, and fleas can invade vulnerable bodies and beds, verminous people can undermine the dominant culture—openly or insidiously. In the United Kingdom, an anonymous sign demanded, "Leave the E. U.: No more Polish vermin," and a Tennessee state legislator, upon hearing that the children born in the United States to Hispanic mothers are entitled to health care, remarked, "Well, they can go out there like rats and multiply, then, I guess."[8] And in 2013, Ken Cuccinelli, Virginia attorney general and a Republican gubernatorial nominee, compared immigration policy to pest control in Washington, DC—which, he charged, was dumping rats across the border into Virginia: "So, anyway, it is worse than our immigration policy. . . . You can't break up rat families. Or raccoons, and all the rest, and you can't even kill 'em. It's unbelievable."[9]

One need not look too hard to find the term vermin applied to feminists who, their enemies believe, will overturn women's traditional roles in society. The same danger arises from other vermin—African Americans, Hispanics, and Eastern Europeans—swarming in, multiplying, and destroying the rightful order.[10] The belief that all these groups must be policed, controlled, and compelled underlies populist and nativist ideologies throughout the world. Those hurling the label of vermin are invoking, consciously or uncon-

sciously, centuries-old motifs about the verminous and the need for extermination.

One thing remains a constant: bugs, rats, and verminous people loom large
in the imagination of those who fear them. Even if they exist only at the
periphery of vision, they are envisioned everywhere—as both metaphor and
literal life forms. Vermin are a frightening metaphor because they can be a
frightening reality. In the past, people believed in the lousy disease—now
they think that every piece of lint in a child's hair is a louse. When a tiny bug
moves across the page of a book (that happened to me yesterday), bedbugs
seem to have crawled into bed with you. And the vociferous scratching of
dogs and cats rings the vermin bell in one's head: Is it fleas, and how should
one respond?

Exterminators have answered this bell from the moment John Southall
started his business in the eighteenth century, becoming the first of many
capitalists and colonialists seeking to profit from vermin. In the nineteenth
century, naturalists and entomologists tried to show the public that the creatures they feared could reward study, and that the myths associated with
them could be dispelled, such as the belief that different races have different
kinds of lice. By the twentieth century, the commodification of vermin produced new kinds of insecticides and rodenticides, some turning out to be as
dangerous to human health—whether in a concentration camp or in a child's
bedroom—as the previously applied mercury and sulphur. The twentieth century also saw attacks on vermin evolve from a private enterprise to a public
policy, whether to help soldiers tormented by lice in World War I trenches or
to reassure residents of cities even more crowded than they seemed.

As cleanliness became a moral condition, the verminous were increasingly ostracized from polite society. It became a belief—a reassuring belief—
that only the poor, the foreign, and the homeless had lice and fleas, and only
their habitations had rats. Consequently, whether one belonged to the newly
risen bourgeoisie of early modern England or North America, or lived in the
gentrifying urban neighborhoods of the twenty-first century, protecting one's
body or home became an urgent concern, for people vulnerable at any moment to a lice notice from school or a rat squeak from the garbage. Nice people do not have un-nice vermin.

But the pretensions of people who consider themselves too good or too
clean to have vermin can be challenged by turning the verminous image
against them. As early as the sixteenth century, the humanist Heinsius saw
this potential when he followed the louse down and up the social ladder, a

"If this is too small I also have a nice German shepherd that just came on the market."

Dog Cartoon #3133, © Mark Anderson. Courtesy of the artist, www.andertoons.com

literary motif extending into early modern times and used in both poetry and pornography. The satirist Jonathan Swift weaponized insects and rats in parodying British society and politics in *Gulliver's Travels*. By the nineteenth century, caricaturists and fairy-tale writers launched pungent attacks using vermin, while scientists used the insects as a way to build a structure of race, and entrepreneurs utilized fleas and rats to entertain. In recent years, it is almost impossible to open a magazine without encountering a cartoon about vermin, aimed at targets like the real estate market.

Recent cartoons—especially about fleas, the insect world's traditional comic relief—offer a cuddlier image of vermin. Sympathetic literature, especially meant for children, has offered white rats as protagonists, reflecting today's sympathy for the natural world. The animal rights organization PETA, in a plea to treat rats nicely, argues, "Rats love to be tickled, and they make chirping noises that sound like laughter." PETA also supports the rights of bugs: "All animals have feelings and have a right to live free from unnecessary suffering—regardless of whether they are considered 'pests' or 'ugly.' "[11]

Today, postmodern sensibilities have rehabilitated the image of many creatures. Animals once considered terrifying vermin are now viewed as noble and majestic, broadly protected and encouraged as endangered species. Perhaps an attitude of tolerance and sympathy with vermin is on its way, especially as the world seems imperiled by climate change. A recent piece in *Discover Magazine* about decreased insect populations warns, "Insects, the most abundant and diverse animals on Earth, are facing a crisis of epic proportions, according to a growing body of research and a rash of alarmist media reports that have followed. If left unchecked, some scientists say, recent population declines could one day lead to a world without insects."[12]

Perhaps we need not be too concerned about bedbugs and lice, which have already found ways to evolve around threats to their existence, or about rats, which can survive nuclear attacks. Whatever their destiny, vermin are likely to endure as a language for condemning other human beings, notably those whom the powerful despise—or fear. Vermin, piercing our skin and attacking our homes, provide a rich image to fling at the other person, culture, or society. To determine which group or individual is considered most threatening at any moment, we can simply track the trail of verminous language. Bedbugs, lice, fleas, and rats have a perverse power over humans, extending beyond any actual threat to our health or homes. We think in terms of animals, and—often with horrific effect—these animals allow us to think the worst.

Introduction • *Vermin in History*

1. Hans Zinsser, *Rats, Lice and History* (New York: Little, Brown, 1935; repr. New York: Black Dog and Leventhal Publishing, 1963), 9.

2. In this book, I concentrate on Great Britain and North America. Many of the same themes are discussed in a more European context, particularly emphasizing the role of vermin in French culture, in Frank Collard and Évelyne Samama, eds., *Poux, Puces, Punaises: La Vermine de l'Homme, Découverte, descriptions et traitements, Antiquité, Moyen Âge, Époque moderne* (Paris: L'Harmattan, 2015). In the introduction, the editors write, "les animaux ont une histoire qui dit beaucoup sur celle de homes. Il n'y avait aucune raison de laisser dans l'ombre poux, puces et punaises au prétexte qu'ils comtent parmi les plus vils et les plus menus" [animals have a history which says much about the history of human beings. There is no reason to leave lice, fleas and bedbugs in the shadows on the pretext that they are considered more vile and more insignificant than man] (7).

3. Historians and philosophers now debate the way history is divided into different chronological periods, particularly when discussing global culture, but for the purposes of this book, I will continue to use the traditional historiography of European history: antiquity, from the beginning of time until the fall of Rome, ca. 476; medieval, 500–1400; the Renaissance, ca. 1400–1600; early modern, 1500–1800; and modern, 1800 to the present. For an interesting discussion of historical and chronological divisions, see Chris Lorenz, " 'The Times They Are a-Changing': On Time, Space and Periodization in History," in *The Palgrave Handbook of Research in Historical Culture and Education*, ed. Mario Garretero, Stefan Berger, and Maria Grever (London: Palgrave Macmillan, 2017), 109–31.

4. For the story of the psychological effect of germ theory, see Nancy Tomes, *The Gospel of Germs: Men, Women, and the Microbe in American Life* (Cambridge, MA: Harvard University Press, 1998).

5. For an enlightening and comprehensive analysis of the emotion of disgust, see William Ian Miller, *The Anatomy of Disgust* (Cambridge, MA: Harvard University Press, 1997). Miller also deconstructs the psychological work of Paul Rozin, particularly his many works on food and disgust. For Rozin's analysis of disgust, see Paul Rozin and April E. Fallon, "A Perspective on Disgust," *Psychological Review* 94 (1987): 23–41.

6. On the varied definitions of vermin, see Lucinda Cole, *Imperfect Creatures: Vermin, Literature, and the Sciences of Life, 1600–1740* (Ann Arbor: University of Michigan Press, 2016), 1; Mary E. Fissell, "Imagining Vermin in Early Modern England," in *The Animal/Human Boundary: Historical Perspectives*, ed. Angela N. H. Creager and William Chester Jordan (Rochester, NY: University of Rochester Press, 2002), 77–114; Karen Raber, *Animal Bodies, Renaissance Culture* (Philadelphia: University of Pennsylvania Press, 2013), 103–25; Harriet Ritvo, *The Animal Estate: The English and Other Creatures in the Victorian Age* (Cambridge, MA: Harvard University Press, 1987), 1–42; Carol Kaesuk Yoon, *Naming Nature: The Clash between Instinct and Science* (New York: W. W. Norton, 2009), 3–22.

7. *The Experienced Vermine-Killer*, in A. S. Gent, *The Husbandman, Farmer and Grasier's Compleat Instructor* (London, 1697), 142.

8. William Robertson, *Ayrshire, Its History and Historic Families* (1908), 2:103, Digitizing Sponsor: National Library of Scotland, https://archive.org/details/ayrshireitshisv21908robe/page/102/mode/2up/search/rats.

9. On the role of skin in culture, see Katharine Young, "Introduction," in *Bodylore*, ed. Katharine Young (Knoxville: University of Tennessee Press, 1993), xviii.

10. Margaret Shildrick, *Embodying the Monster: Encounters with the Vulnerable Self* (London: Sage, 2002), defines vulnerability as "an existential state that may belong to any of us, but which is characterized nonetheless as a negative attribute, a failure of self-protection, that opens the self to the potential of harm" (1). Shildrick is discussing monsters and the threat they pose to the "normative embodied self," but her analysis translates well into a discussion of the way bedbugs threatened the body.

11. Thomas Swaine, *The Universal Directory for Taking Alive, or Destroying, Rats and Mice, by a Method Hitherto Unattempted, etc., etc.* (1783), ii–iii.

12. Henry Mayhew, *London Labour and the London Poor*, 4 vols. (London: Griffin, Bohn, 1861; repr., Dover, 1968), 3:37–38.

13. In early modern times, the Gypsies were seen as a group who originated in Southern Europe or Bohemia; some writers thought they were distinguished by dark or yellow skin. They were generally viewed negatively. The *Oxford English Dictionary* defines the earlier meaning of "Gypsy" as "A person who possesses qualities or characteristics supposed to be typical of Gypsies; *spec.* †(*a*) a person who acts in a disreputable, unscrupulous, or deceptive manner (*obsolete*)." By the twentieth century, a racialized understanding of "Gypsy" was added to this description and used to justify extermination by the Nazis.

14. Though dated, the place to start in any examination of appropriate behavior is Norbert Elias, *The Civilizing Process*, vol. 1, *The History of Manners*, trans. Edmund Jephcott (New York: Pantheon Books, 1978); originally published as *Über den Prozess der Zivilisation* (Zurich: Haus zum Falken, 1939). For a more recent discussion, which approaches the body and cleanliness within the context of race and gender discourses, see Kathleen A. Brown, *Foul Bodies: Cleanliness in Early America* (New Haven, CT: Yale University Press, 2009). See also Edward Muir, *Ritual in Early Modern Europe* (Cambridge: Cambridge University Press, 2005), 125–54. On eighteenth-century unease with the body, see Carol Houlihan Flynn, *The Body in Swift and Defoe* (Cambridge: Cambridge University Press, 1990).

15. For the changes in attitudes toward cleanliness, based on an analysis of what

was considered noxious to the senses in the seventeenth and eighteenth centuries, see Emily Cockayne, *Filth, Noise, and Stench in England, 1600–1770* (New Haven, CT: Yale University Press, 2007).

16. Steven Johnson, *The Ghost Map: The Story of London's Most Terrifying Epidemic— And How It Changed Science, Cities, and the Modern World* (London: Riverhead Books, 2006), describes how the architect John Nash in the nineteenth century designed Regent Street as "a kind of *cordon sanitaire* separating the well-to-do Mayfair from the growing working-class community of Soho" (20).

17. "Duke Street Area: Duke Street, West Side," in *Survey of London*, vol. 40, *The Grosvenor Estate in Mayfair, Part 2 (The Buildings)*, ed. F. H. Sheppard (London: London City Council, 1980), 91–92, https://www.british-history.ac.uk/survey-london/vol40/pt2/pp91-92.

18. On the cultural meaning of smell and the growing repugnance that certain odors generated among the upper classes, see Alain Corbin, *The Foul and the Fragrant: Odor and the French Social Imagination* (Cambridge, MA: Harvard University Press, 1986), and Constance Classen, David Howes, and Anthony Synnott, *Aroma: The Cultural History of Smell* (London: Routledge, 1994). Classen, Howes, and Synnott comment that "smell has been marginalized because it is felt to threaten the abstract and impersonal regime of modernity by virtue of its radical interiority, its boundary-transgressing propensities and its emotional potencies" (5).

19. George Washington, *The Rules of Civility: The 110 Precepts That Guided Our First President in War and Peace*, ed. George Brookhiser (New York: Free Press, 1997), 30.

20. Benjamin Martin, *The Young Gentleman and Lady's Philosophy in a Continued Survey of the Works of Nature* (London: W. Owen, 1782), 3:61.

21. Mayhew, *London Labour and the London Poor*, 3:19.

22. "Who Is the Savage?," New Perspectives on the West, PBS Interactive, https://www.pbs.org/weta/thewest/program/episodes/four/whois.htm.

23. "Against Insects," *Popular Mechanics* 81 (1944): 67.

24. Quoted in Erin Steuter and Deborah Wills, *At War with Metaphor: Media, Propaganda, and Racism in the War on Terror* (Lanham, MD: Lexington Books, 2008), 122.

25. Andrew Jacobs, "Just Try to Sleep Tight. The Bedbugs are Back," *New York Times*, November 27, 2005, Section 1, 1.

26. Toluse Olorunnipa, Daniel A. Fahrenthold, and Jonathan O'Connell, "Trump Has Awarded the Next Year's G-7 Summit of World Leaders to His Miami-Area Resort," *Washington Post*, October 17, 2019, https://www.washingtonpost.com/politics/trump-has-awarded-next-years-g-7-summit-of-world-leaders-to-his-miami-area-resort-the-white-house-said/2019/10/17/221b32d6-ef52-11e9-89eb-ec56cd414732_story.html.

27. Lorraine Daston and Gregg Mitman, "Introduction," in *Thinking with Animals: New Perspectives on Anthropomorphism*, ed. Lorraine Daston and Gregg Mitman (New York: Columbia University Press), 2.

28. Giovanni Boccaccio, *The Decameron of Giovanni Boccaccio*, ed. Henry Morley (London: Routledge, 1895), 12.

Chapter 1 • **"That *Nauseous* Venomous Insect"**

1. Samuel Pepys, *The Diary of Samuel Pepys*, 10 vols., ed. Robert Latham and William Matthews (Berkeley: University of California Press, 1974), April 23, 1662, 3:70;

June 11, 1668, 9:231. Dr. Clerke (aka Clarke) was a physician to Charles II and a fellow member of the Royal Society.

2. On the recent historiographical treatment of the body, see Roy Porter, "History of the Body," in *New Perspectives on Historical Writing*, ed. Peter Burke (University Park: Pennsylvania State University Press, 1991), 206–26; and Peter Burke, *What Is Cultural History?*, 2nd ed. (Cambridge: Polity, 2008). See also Caroline Walker Bynum's critique of feminist and poststructuralist analyses of the body as a discursive construct, "Why All the Fuss about the Body: A Medievalist's Perspective," *Critical Inquiry* 22 (Autumn 1995): 1–33.

3. Oliver Goldsmith, *An History of the Earth and Animated Nature*, 2nd ed., 8 vols. (London: J. Nourse, 1779), 2:281–82. The first edition of this work appeared in 1775. A plagiarized version of this description also appears in William Frederic Martyn, *A New Dictionary of Natural History; or Compleat Universal Display of Animate Nature* (London: Harrison, 1785).

4. On phenomenology and the body, see Thomas J. Csordas, "Embodiment as a Paradigm for Anthropology," *Ethos* 18 (1990): 5–47.

5. Bynum, "Why All the Fuss," 3.

6. On the body as a cultural construct, see the various essays in Christopher Lawrence and Steven Shapin, eds., *Science Incarnate: Historical Embodiments of Natural Knowledge* (Chicago: University of Chicago Press, 1998); and Roy Porter, "The History of the Body Reconsidered," in *New Perspectives on Historical Writing* (London: Polity Press, 1991), 233–60.

7. In the seventeenth century, the French philosopher René Descartes argued that rationality and the body were completely different and that only human beings possessed a soul or reason. Not many people suffering from bedbugs in early modern times would have agreed with him. Many historians, philosophers, and anthropologists view Cartesian dualism as the starting point of the divorce between the body, nature, and the soul. See Steven Shapin, "The Philosopher and the Chicken: On the Dietetics of Disembodied Knowledge," in *Science Incarnate: Historical Embodiments of Natural Knowledge*, ed. Christopher Lawrence and Steven Shapin (Chicago: University of Chicago Press, 1998), 21–49.

8. John Southall, *A Treatise of Buggs* (London: J. Roberts, 1730), 2–3.

9. L. O. J. Boynton, "The Bed-Bug and the 'Age of Elegance,'" *Furniture History* 1 (1965): 15–31, 16–17.

10. Thomas Muffet, *The Theater of Insects*, vol. 3 of *The History of Four-Footed Beasts Serpents and Insects* (London: Printed by E. C., 1658; repr., New York: Da Capo Press, 1967), 3:1096–97. According to the modern entomologist Richard J. Pollack, bedbugs smell like coriander but are usually undetectable unless there is a large infestation. He writes, "The perception is in the nose of the beholder, indeed" (personal email to author, May 4, 2011). I wish to thank Professor Pollack for his unique perspective on this question. The ancients also associated the smell of bedbugs with coriander, which presumably they did not dislike as much as the moderns; see Michael F. Potter, "Bed Bug History," *American Entomologist* 57 (2011): 14–32.

11. John Ray, *Observations Topographical, Moral, & Physiological; Made in a Journey through Part of the Low-Countries, Germany, Italy, and France* (London: John Martyn, 1673), 411.

12. "Bugbear," *Oxford English Dictionary Online*, accessed October 11, 2020, http://www.oed.com.ezproxy.proxy.library.oregonstate.edu/view/Entry/24351?isAdvanced=false&result=1&rskey=LbYJJm&.

13. Here I follow the analytical framework of the cultural historian Robert Darnton, who writes in his classic *The Great Cat Massacre and Other Episodes in French Cultural History* (New York: Vintage Books, 1984) that "When we cannot get a joke, or a ritual, or a poem, we know we are on to something. . . . We may be able to unravel an alien system of meaning" (5).

14. J. R. Busvine, *Insects, Hygiene and History* (London: Athlone Press, 1976), 30–34. Any historian of parasitical insects must acknowledge her debt to Busvine, whose encyclopedic and engaging book is the starting point for any research on this subject. L. O. J. Boynton's "The Bed-Bug and the 'Age of Elegance' " is also invaluable for its insights into beds and bugs. Two recent books explore the subject of verminous insects: J. F. M. Clark, *Bugs and the Victorians* (New Haven, CT: Yale University Press, 2009); and Amy Stewart, *Wicked Bugs: The Louse That Conquered Napoleon's Army and Other Diabolical Insects* (Chapel Hill, NC: Algonquin Books, 2011). A note on the spelling of "bedbug": in the eighteenth century, bedbugs were generally referred to as "bugs," rather than bedbugs. Thus almost every reference to "the bug" during this period indicates the *Cimex lectularius*, or the bedbug.

15. Goldsmith, *History of the Earth and Animated Nature*, 2:281.

16. George Adams, *Essays on the Microscope* (London: Robert Hindmarsh, 1787), 698, http://find.galegroup.com/bncn/start.do?prodId=BBCN&userGroupName=uclosangeles.

17. Louis Lémery, *A Treatise of All Sorts of Food, Both Animal and Vegetable; Also of Drinkables* (London, 1745), 157.

18. Alain Corbin, *The Foul and the Fragrant: Odor and the French Social Imagination* (Cambridge, MA: Harvard University Press, 1986), 1–8; and Constance Classen, David Howes and Anthony Synnott, *Aroma: The Cultural History of Smell* (New York: Routledge, 1994), 51–66.

19. On the humorous aspects of lice and fleas, see Busvine, *Insects, Hygiene and History*, 76–79.

20. Thomas Tryon, *A Treatise of Cleaness in Meats and Drinks of the Preparation of Food* (London, 1682), 7–8, http://find.galegroup.com/bncn/start.do?prodId=BBCN&userGroupName=uclosangeles.

21. Antonie van Leeuwenhoek and Francesco Redi had shown that Aristotle's account of the spontaneous generation of insects was incorrect by 1677.

22. Mark Jenner, "The Politics of London Air: John Evelyn's *Fumifugium* and the Restoration," *Historical Journal* 38 (1995): 536–51.

23. Thomas Tryon, *A Way to Health, Long Life and Happiness* (1691), 440.

24. Peter Earle, *The Making of the English Middle Class: Business, Society and Family Life in London, 1660–1730* (Berkeley: University of California Press, 1989), 13. For a description of the people who constituted this class, see 3–16.

25. On the increasing sensitivity to smell, see Katherine Ashenburg, *The Dirt on Clean: An Unsanitized History* (New York: North Point Press, 2007), 146; and Kathleen M. Brown, *Foul Bodies* (New Haven, CT: Yale University Press, 2009), 243–46.

26. Edward Ward, *The History of the London Clubs, Part I* (London: J. Dutton, 1709), 19.

27. Tryon, *Way to Health*, 434.

28. Richard Mead, *A Short Disclosure Concerning Pestilential Contagion, and the Methods Used to Prevent It* (London: Sam. Buckley, 1720), 48. Virginia Smith discusses Mead and other proponents of cleanliness in *Clean: A History of Personal Hygiene and Purity* (Oxford: Oxford University Press, 2007), 220–28.

29. *Read's Weekly Journal or British Gazetteer*, No. 5074, August 16, 1760. All eighteenth-century newspaper citations come from the Seventeenth and Eighteenth-Century Burney Newspapers Collection, http://find.galegroup.com/bncn/start.do ?prodId=BBCN&userGroupName=uclosangeles.

30. *London Chronicle*, No. 496, February 25, 1760 .

31. *St. James Evening Post*, No. 2862, October 9, 1733.

32. *Daily Courant*, No. 5356, June 7, 1733.

33. *Daily Gazetteer*, No. 298, June 10, 1736.

34. *Whitehall Evening Post or London Intelligence*, No. 503, May 13, 1749.

35. On beds and the middle classes, see Earle, *Making of the English Middle Class*, 291–93; Doreen Yarwood, *The English Home* (London: B. T. Batsford, 1979), 116, 134; and Lawrence Wright, *Warm and Snug: The History of the Bed* (London: Routledge & Kegan Paul, 1962).

36. William Cauty, *Natura, Philosophia, and Ars in Concordia. Or, Nature, Philosophy, and Art in Friendship* (London, 1772), 82, https://wellcomecollection.org/works/ w8qjfzku/items?canvas=1&langCode=eng&sierraId=b30376695.

37. Tryon, *Way to Health*, 442–43.

38. Southall, *Treatise of Buggs*, 16–17.

39. Cauty, *Natura, Philosophia, and Ars in Concordia*, 79.

40. "Report from August 17, 1751," British National Archives Navy Board Records, Adm 106/1093/348. Scientists now know that this irritating skin condition is actually caused by the itch mite, not the bedbug.

41. *Royal London Evening Post*, No. 1618, March 28, 1738. The political and propaganda uses of the *London Evening Post* are discussed in G. A. Cranfield, "The 'London Evening Post,' 1727–1744: A Study in the Development of the Political Press," *Historical Journal* 6 (1963): 1:20–37, http://www.jstor.org/stable/3020548.

42. Goldsmith, *History of the Earth and Animated Nature*, 282.

43. Martyn, *New Dictionary of Natural History*; Francis Fitzgerald, *The General Genteel Preceptor by Francis Fitzgerald*, 2nd ed. (London: C. Taylor, 1797), 1, http://find .galegroup.com/bncn/start.do?prodId=BBCN&userGroupName=uclosangeles.

44. Cauty, *Natura, Philosophia, and Ars in Concordia*, 84–85.

45. Quoted in Anthony Burgess and Francis Haskell, *The Age of the Grand Tour* (New York: Crown, 1967), 39.

46. *The Universal Family-Book: or, A Necessary and Profitable Companion for All Degrees of People of Either Sex* (1703), 197–98.

47. Boyle Godfrey, *Miscellanea Vere Utile; or Miscellaneous Experiments and Observations on Various Subjects* (London, 1735?), 134.

48. Noel Chomel, *Dictionaire* [sic] *Oeconomique: or the Family Dictionary. Containing the Most Experience'd Methods of Improving Estates and of Preserving Health*, trans. R. Bradley (1727); Godfrey, *Miscellanea Vere Utile*, 131.

49. Chomel, *Dictionaire Oeconomique*.

50. Cauty, *Natura, Philosophia, and Ars in Concordia*, 81.

51. Eric H. Ash, ed., "Introduction," in "Expertise: Practical Knowledge and the Early Modern State," ed. Eric H. Ash, special issue, *Osiris* 25 (2010), defines an expert: "To be 'expert' was to possess and control a body of specialized practical and productive knowledge, not readily available to everyone" (5). Ash explains that this is a "provisional" definition and argues that expertise is a socially constructed category that depends on many factors, including legitimization by social status and institutional support.

52. On quackery and medicine, see the many works of Roy Porter, especially *Health for Sale: Quackery in England 1660–1850* (Manchester: Manchester University Press, 1989).

53. *Daily Journal*, No. 4256, September 17, 1734.

54. *General Advertiser*, No. 5527, July 8, 1752.

55. *Public Ledger*, No. 387, October 7, 1761.

56. Cauty, *Natura, Philosophia, and Ars in Concordia*, iv.

57. *British Magazine and Review; or, Universal Miscellany*, vol. 3 (1783), 352–53.

58. Cauty, *Natura, Philosophia, and Ars in Concordia*, 84.

59. Patrick Browne, *The Civil and Natural History of Jamaica* (London, 1789), 434, http://find.galegroup.com/bncn/start.do?prodId=BBCN&userGroupName=uclosangeles.

60. John Southall, *Treatise of Buggs*, 2.

61. On the perils of the New World for English colonizers and their corresponding obsession with the deprecations of the skin, see Emily Senior, "'Perfectly Whole': Skin and Text in John Gabriel Stedman's *Narrative of a Five Years Expedition against the Revolted Negroes of Surinam*," *Eighteenth-Century Studies* 44, no. 1 (2010): 39–56.

62. *Lloyd's Evening Post*, July 15, 1768.

63. Southall, *Treatise of Buggs*, 14.

64. I borrow the term "African Magi" from Susan Scott Parrish, *American Curiosity: Cultures of Natural History in the British Colonial Atlantic World* (Chapel Hill: University of North Carolina Press, 2006), 247.

65. Hans Sloane, *A Voyage to the Islands Madera, Barbados, Nieves, S. Christophers and Jamaica*, 2 vols. (1707); John Woodward, *An Essay toward a Natural History of the Earth and Terrestrial Bodies, Especially Minerals, etc.* (London, 1695); *Brief Instructions for Making Observations in All Parts of the World* (London, 1696); and *An Attempt towards a Natural History of the Fossils of England* (1728, 1729).

66. Journal Book Copy XIII (1726–31), Archives of the Royal Society, London; Southall, *A Treatise of Buggs*, x–xi.

67. Southall, *Treatise of Buggs*, 2.

68. *London Daily Post*, March 15, 1740.

69. Southall, *Treatise of Buggs*, 22–23.

70. Southall, *Treatise of Buggs*, 24.

71. Southall, *Treatise of Buggs*, 25.

72. Southall, *Treatise of Buggs*, 27.

73. Southall, *Treatise of Buggs*, 31.

74. Southall, *Treatise of Buggs*, 29.

75. Southall, *Treatise of Buggs*, 38–39.

76. Southall, *Treatise of Buggs*, 39.

77. His will can be found in the National Archives, London, J90/1049.

78. J. Southall, *A Treatise on the Cimex Lectularius* (1793), 43–46.

79. Busvine, *Insects, Hygiene and History*, 85.

80. *Public Advertiser*, September 3, 1763.

81. Samuel Sharp, *Letters from Italy, Describing the Customs and Manners of That Country, in the Years 1765 and 1766*, 2nd ed. (1767), 239.

82. On the use of animals in the construction of culture and the changes between pre-Cartesian and post-Cartesian understandings of their meanings, see Erica Fudge, *Brutal Reasoning: Animals, Rationality, and Humanity in Early Modern England* (Ithaca, NY: Cornell University Press, 2006), 175–93. Fudge stresses the agency of animals in the early modern world and their integration to the point of union with humans prior to Descartes's mechanization and silencing of animals (190). The classic account of animals in the premodern West is Keith Thomas, *Man and the Natural World: A History of the Modern Sensibility* (New York: Pantheon Books, 1983).

Chapter 2 • Bedbugs Creeping through Modern Times

1. "Impact of Bed Bugs More Than Skin Deep," *Medscape*, May 16, 2011, https://www.medscape.com/viewarticle/74277/.

2. There are many accounts of the psychological harm wrought by bedbugs, including an entire book inspired by the horror they evoke: Brooke Borel, *Infested: How the Bed Bug Infiltrated Our Bedrooms and Took Over the World* (Chicago: University of Chicago Press, 2015). Also, see "Bed-Bug Madness: The Psychological Toll of the Blood Suckers, *The Atlantic*, October 16, 2014, https://www.theatlantic.com/health/archive/2014/10/bed-bug-madness-the-psychological-toll-of-the-blood-suckers/381447/; "Teen Trying to Kill Bed Bug Causes $300,00 in Fire Damage in Cincinnati," *New York Post*, November 30, 2017, https://nypost.com/2017/11/30/teen-trying-to-kill-bed-bug-causes-300k-fire/.

3. "Six Facts You Didn't Know about Bed Bugs," Pest World, https://www.pestworld.org/news-hub/pest-articles/six-facts-you-didnt-know-about-bed-bugs/.

4. William Quarles, "Dispersal a Consequence of Bed Bug Biology," *IPM Practitioner* (March/April 2010): 32.

5. Jessica Goldstein, "An Army of Bedbugs Were Partying in My Bed," *Washington Post*, March 13, 2014, https://www.washingtonpost.com/lifestyle/magazine/an-army-of-bedbugs-was-partying-in-her-mattress-what-else-could-go-wrong/2014/03/13/318fd23e-8dca-11e3-98ab-fe5228217bd1_story.html?noredirect=on&utm_term=.ee68d130a5f8.

6. William Bingley, *Animal Biography: Or Popular Zoology*, 4 vols. (London: F. C. and J. Rivington, 1820), 3:70, https://www.google.com/books/edition/Animal_biography_or_Popular_zoology/CzMTAAAAQAAJ?hl=en&gbpv=1&bsq=bedbug .https://www.google.com/books/edition/Animal_biography_or_Popular_zoology/CzMTAAAAQAAJ?hl=en&gbpv=1&bsq=bedbug.

7. Jane Carlyle to Thomas Carlyle, September 13, 1852, quoted in Judith Flanders, *Inside the Victorian Home: A Portrait of Domestic Life in Victorian England* (New York: W. W. Norton, 2003), 49.

8. On the establishment of the Entomological Society of London in 1833, see J. F. M. Clark, *Bugs and the Victorians* (New Haven, CT: Yale University Press, 2009), 10–11.

9. "Sixth Annual Fair of the American Institute of New York," in *Mechanics' Magazine and Journal of the Mechanical Institute* (New York: D. E. Minor, 1834), 158.

10. For the history of the pest control industry in the United States, see Robert Snetsinger, *The Ratcatcher's Child: The History of the Pest Control Industry* (Cleveland, OH: Franzak & Foster, 1983).

11. Katherine Butler, "Scientists Discover New Weapon in Fight against Bedbugs," *Mother Nature Network*, September 9, 2010, https://www.mnn.com/earth-matters /animals/stories/scientists-discover-new-weapon-in-fight-against-bedbugs.

12. "Orkin Releases Top 50 Bed Bug Cities List," Orkin, January 8, 2018, https:// www.orkin.com/press-room/orkin-releases-top-50-bed-bug-cities-list/.

13. Marshall Sella, "Bedbugs in the Duvet," *New York Magazine*, May 2, 2010, http:// nymag.com/news/features/65733/.

14. Donald McNeill, "They Crawl, They Bite, They Baffle Scientists," *New York Times*, August 30, 2010, https://www.nytimes.com/2010/08/31/science/31bedbug.html. One can look up all of the articles on bedbugs published by *New York Times* online at https://www.nytimes.com/topic/subject/bedbugs; see also Borel, *Infested*; and Jeffrey A. Lockwood, *The Infested Mind: Why Humans Fear, Loathe, and Love Insects* (Oxford: Oxford University Press, 2013), 186–88.

15. See, e.g., "Recognising Bed Bugs and Preventing Infestation," Gouvernment du Québec, June 15, 2018, https://www.quebec.ca/en/homes-and-housing/healthy-living -environment/recognising-bed-bugs-and-preventing-infestation/; and Benedict Moore-Bridger, "London Is Being Infested by Super Resistant Bed Bugs," *Evening Standard*, September 29, 2016, https://www.standard.co.uk/news/london/revealed -how-new-superresistant-bedbugs-are-infesting-london-a3356811.html.

16. Andrew Jacobs, "Just Try to Sleep Tight: The Bedbugs Are Back," *New York Times*, November 27, 2005, https://www.nytimes.com/2005/11/27/nyregion/just-try -to-sleep-tight-the-bedbugs-are-back.html.

17. Emily B. Hager, "What Spreads Faster than Bedbugs? Stigma," *New York Times*, August 20, 2010, https://www.nytimes.com/2010/08/21/nyregion/21bedbugs.html.

18. There seems to be some recent debate among entomologists about the possibility that bedbugs transmit pathogens that cause Chagas disease, which has caused the deaths of thousands of people in South America. See "Study Offers Further Evidence of Bed Bugs' Ability to Transmit Chagas Disease Pathogen," *Entomology Today*, January 30, 2018, https://entomologytoday.org/2018/01/30/.

19. Jane E. Brody, "Keeping Those Bed Bugs from Biting," *New York Times*, April 13, 2009, https://www.nytimes.com/2009/04/14/health/14brod.html; Albert C. Yan, "Bedbugs, Scabies and Head Lice—Oh, My," American Academy of Dermatologists, March 4, 2010, https://www.aad.org/media/news-releases/bedbugs-scabies-and-head-lice-oh-my.

20. M. Jane Pritchard and Stephen W. Hwang, "Severe Anemia from Bedbugs," *Canadian Medical Association Journal* 18, no. 5 (2009): 287–88, https://www.ncbi.nlm .nih.gov/pmc/articles/PMC2734207/.

21. Tess Russell, "Alone When the Bedbugs Bite," *New York Times*, November 11, 2010, https://www.nytimes.com/2010/11/21/fashion/21Modern.html.

22. Amy Schumer, *The Girl with the Lower Back Tattoo* (New York: Simon and Schuster, 2017), 245.

23. William Kirby and William Spence, *An Introduction to Entomology: Or Elements*

of the Natural History of Insects, vol. 1, 5th ed. (London: Longman, Rees, Orme, Brown, and Green, 1828), 107.

24. On Kirby and the role of theology in the nineteenth-century naturalist tradition, see Clark, *Bugs and the Victorians*, 44–53, and Paul L. Farber, *Finding Order in Nature* (Baltimore: Johns Hopkins University Press, 2000), 6–21.

25. Carl van Linné, *A General System of Nature, through the Three Grand Kingdoms of Animals, Vegetables, and Minerals*, vol. 2, trans. William Turton (1762–1835) (London: Lackington, Allen and Co.,1806), 608.

26. John Mason Good, Olinthus Gregory, and Newton Bosworth, *Pantologia: A New Cabinet Cyclopaedia, Comprehending a Complete Series of Essays, Treatises, and Sermons* (Edinburgh: J. Walker, 1819).

27. Thaddeus William Harris, "A Report on the Insects of Massachusetts, Injurious to Vegetation," Biodiversity Heritage Library, https://www.biodiversitylibrary.org/bibliography/6091.

28. Stephen A. Kells and Jeff Hahn, "Prevention and Control of Bed Bugs in Homes," University of Minnesota Extension, last reviewed in 2018, https://extension.umn.edu/biting-insects-and-insect-relatives/bed-bugs.

29. J. R. Busvine, *Insects, Hygiene and History* (London: Athlone Press, 1976), 82–85.

30. Stephen L. Doggett, Dominic E. Dwyer, Pablo F. Peñas, and Richard C. Russell, "Bed Bugs: Clinical Relevance and Control Options," *Clinical Microbiology Review* 25, no. 1 (2012): 164–92, http://cmr.asm.org/content/25/1/164.short.

31. Busvine, *Insects, Hygiene and History*, 85.

32. Cara Buckley, "Doubts Rise on Bedbug-Sniffing Dogs," *New York Times*, November 11, 2010, https://www.nytimes.com/2010/11/12/nyregion/12bedbugs.html.

33. "BedBugs 101," AppAdvice.com, https://appadvice.com/app/bedbugs-101/411932955.

34. Joshua B. Benoit, Seth A. Phillips, Travis J. Croxall, Brady Chrisensen, Jay A. Yoder, and David L. Denlinger, "Addition of Alarm Pheromone Components Improves the Effectiveness of Desiccant Dusts against *Cimex lectularius*," *Journal of Medical Entomology* 46, no. 3 (2009): 572–79. See also Anders Aak, Espen Roligheten, Bjørn Arne Rukke, and Tone Birkemoe, "Desiccant Dust and the Use of CO_2 Gas as a Mobility Stimulant for Bed Bugs: A Potential Control Solution?," *Journal of Pest Science* 90, no. 1 (2017): 249–59.

35. "Bed Bug Summits," US Environmental Protection Agency, last updated October 11, 2016, https://archive.epa.gov/epa/bedbugs/bed-bug-summits.html.

36. First Annual National Bed Bug Summit, US Environmental Protection Agency, Arlington, VA, April 14–15, 2009; "EPA Co-Hosts National Bed Bug Summit to Address the Return of a Pest," *Science Matters Newsletter* (April 2011); and Federal Bed Bug Workgroup, *Collaborative Strategy on Bed Bugs* (Washington, DC: Environmental Protection Agency, February 2015), https://www.epa.gov/sites/production/files/2015-02/documents/fed-strategy-bedbug-2015.pdf.

37. Henry Mayhew (1812–87) was the author of *London Labour and the London Poor*, 4 vols. (London: Griffin, Bohn, 1861; repr., Dover, 1968). It consists of a collection of newspaper pieces detailing almost every detail of lower-class society.

38. Mayhew, *London Labour and London Poor*, 3:38.

39. Mayhew, *London Labour and London Poor*, 3:37.

40. Mayhew, *London Labour and London Poor*, 3:39.

41. Mayhew, *London Labour and London Poor*, 3:37–38.

42. Clark, *Bugs and the Victorians*, 8, 12.

43. Beatrix Potter, March 14, 1883, quoted in Flanders, *Inside the Victorian Home*, 48. To combat the bedbugs, Potter used Keating Powder, an insecticide that was still popular during World War I.

44. John S. Farmer and W. E. Henley, eds., *Slang and Its Analogues: Past and Present. A Dictionary of the Heterodox of Society of All Classes of Society for More Than Three Hundred Years*, vol. 5 (London: Harrison and Sons, 1902), 65; Richard Jones, "Norfolk Howard," Jack the Ripper Tour, accessed October 6, 2019, www.jack-the-ripper-tour .com.

45. Jane W. Carlyle to Thomas Carlyle, August 18, 1843, The Carlyle Letters Online, Center for Digital Humanities, University of South Carolina, http://carlyleletters.duke upress.edu/.

46. C. L. Marlett, "The Bedbug" (1916), quoted in Michael Potter, "The History of Bed Bug Management," *American Entomologist* 57, no. 1 (2011): 14–25, https://academic .oup.com/ae/article/57/1/14/2462090.

47. See, e.g., the fine levied against the Western Exterminator Company in 1913: US Department of Agriculture, "Notice of Insecticide Act Judgment No. 53," Hathi Trust Digital Library, accessed October 7, 2019, https://babel.hathitrust.org/cgi/pt?id=coo .31924055534881&view=1up&seq=117.

48. C. Killick Millard, "Presidential Address on an Unsavory but Important Feature of the Slum Problem (1932)," quoted in Potter, "History of Bed Bug Management," 16.

49. Busvine, *Insects, Hygiene and History*, 63.

50. Medical Research Council, *Report of the Committee on Bed-Bug Infestation 1935–40* (London: His Majesty's Stationary Office, 1942), 5, 37.

51. William C. Gunn, "Domestic Hygiene in the Prevention and Control of Bed-Bug Infestation in Privy Council," in Medical Research Council, *Report of the Committee on Bed-Bug Infestation*, 36, 39.

52. Medical Research Council, *Report of the Committee on Bed-Bug Infestation*, 37.

53. Medical Research Council, *Report of the Committee on Bed-Bug Infestation*, A2.

54. Michael Potter, "The Perfect Storm: An Extension View on Bed Bugs," *American Entomologist* 52, no. 1 (April 1, 2016): 102–4.

55. National Pest Management Association, "2011 Bugs without Borders Survey: New Data Shows Bed Bug Pandemic Is Growing," Pestworld.org, August 17, 2011, https://www.pestworld.org/news-hub/press-releases/2011-bugs-without-borders -survey-new-data-shows-bed-bug-pandemic-is-growing/.

56. Rachel Feltman, "How'd the Bedbug Get Its Bite? Scientists Look to Its Genome for Clues," *Washington Post*, February 2, 2016, https://www.washingtonpost.com/news /speaking-of-science/wp/2016/02/02/howd-the-bedbug-get-its-bite-scientists-look -to-its-genome-for-clues/?utm_term=.5f276e16b5c5.

57. Quoted in Brody, "Keeping Those Bed Bugs from Biting."

58. Butler, "Scientists Discover New Weapon"; Benoit et al., "Addition of Alarm Pheromone."

59. Christopher Terrall Nield, "In Defense of the Bed Bug," The Conversation, February 2, 2016, https://theconversation.com/in-defence-of-the-bed-bug-54218.

60. Warren Booth, "Host Association Drives Genetic Divergence in the Bed Bug,"

Molecular Ecology 24, no. 5 (2015): 980–92, https://www.ncbi.nlm.nih.gov/pubmed /25611460.

61. Jose Lambiet, "Traveler: Bedbugs Devoured Me at Trump Resort," *Miami Herald*, September 7, 2016, https://www.miamiherald.com/entertainment/ent-columns -blogs/jose-lambiet/article100482407.html.

62. Quoted in Eric Grundhauser, "Interviewing the Country's Preeminent Bed Bug Lawyer," Atlas Obscura, June 24, 2016, https://www.atlasobscura.com/articles /interviewing-the-countrys-preeminent-bed-bug-lawyer.

63. Quoted in Erin Fuchs, "Hotels Brace for Next Bite from Bedbug Suits," Law360, December 10, 2010, https://www.law360.com/articles/207424/hotels-brace-for-next -bite-from-bedbug-suits.

64. Quoted in Kate Murphy, "Bedbugs Bad for Bedbugs? Depends on the Business," *New York Times*, September 7, 2010, https://www.nytimes.com/2010/09/08/business /08bedbug.html.

65. "Can I File a Lawsuit against a Hotel or Motel for Bed Bug Bites or Injuries?," Dell and Schaefer Personal Injury Lawyers, accessed November 10, 2020, https://www .dnslaw.com/can-i-file-lawsuit-against-hotel-motel-bed-bug-bites-injuries/.

66. "About the Bed Bug Injury Attorneys," Bed Bug Law, accessed November 10, 2020, https://www.bedbuglaw.com/about-bed-bug-injury-attorneys/.

67. Quoted in Anna Drezen, "The Bizarre and Fascinating World of Bedbug Message Boards," Daily Dot, December 8, 2014, https://www.dailydot.com/unclick/bedbug-tips -tricks-online-forums/.

68. Terramera, "Proof by Terramerra Launches Revolutionary Sprayless Treatment for Bed Bugs," press release, April 10, 2019, https://www.terramera.com/newsroom /proof-by-terramera-launches-revolutionary-sprayless-treatment-for-bed-bugs. The company uses neem oil in its bedbug insecticides, CIRKIL and Proof.

69. "Neem Oil: General Fact Sheet," National Pesticide Information Center, Oregon State University, accessed September 9, 2020, http://npic.orst.edu/factsheets/neemgen .html#whatis.

70. Kate Murphy, "Bedbugs Bad for Business? Depends on the Business," *New York Times*, September 7, 2010, https://www.nytimes.com/2010/09/08/business/08bedbug .html; Cara Buckley, "Doubts Rise on Bedbug-Sniffing Dogs," *New York Times*, November 11, 2010, https://www.nytimes.com/2010/11/12/nyregion/12bedbugs.html.

71. For more about Roscoe, see "Meet Roscoe the Bed Bug Dog: Bell Environmental's Lead Canine Inspector," Bell Environmental Services, accessed November 10, 2020, https://bell-environmental.com/wheres-roscoe/meet-roscoe/.

72. "EPA Co-Hosts National Bed Bug Summit," *Science Matters Newsletter* 2, no. 2 (March/April 2011): https://archive.epa.gov/epa/sciencematters/epa-science-matters -newsletter-volume-2-number-2.html; Moore-Bridger, "London Is Being Infested."

73. Butler, "Scientists Discover New Weapon."

74. Personal correspondence. I wish to thank Dr. Pollack for this information.

75. "Sleep Tight, Starting Tonight," EcoRaider, accessed March 10, 2017, https:// www.amazon.com/EcoRaider—Protection—Non-toxic—Entomological—Publication /dp/B0077CPANQ.

76. "Personnality Insect Pendant Necklace Chic Bedbug Jewelry Long Chain Gift,"

Amazon.com, accessed November 10, 2020, https://www.amazon.com/PONCTUEL
-ESCARGOT-Personnality-Pendant-Necklace/dp/B087ZG9V13.

77. Borel, *Infested*, 134.

78. Bedbugger.com forum, accessed April 11, 2016, http://bedbugger.com/forum
/topic/bed-bug-ptsd-does-it-ever-end; http://bedbugger.com/forum/topic/former
-bed-bug-victims-story-tips-to-overcoming-aniexty-killing-them-and-more.

79. Quoted in Borel, *Infested*, 134.

80. "Alcohol Is a Very Flammable Contact Killer for Bed Bugs," Bedbugger.com,
November 21, 2015.

81. Doggett et al., "Bed Bugs."

82. Feltman, "How'd the Bedbug Get Its Bite?"

83. "Don't Let the Bedbugs Bite," BBC, September 3, 2012, http://www.bbc.co.uk
/news/mobile/magazine-11165108.

84. Mary Wisniewski, "CTA Pulls Red Line Car after Bedbug Report," *Chicago
Tribune*, September 28, 2016, http://www.chicagotribune.com/news/ct-cta-red-line
-bed-bugs-20160927-story.html#.

85. "Lice, Not Bed Bugs Found on Chicago RTA Red Line Train," Bedbugger.com,
September 29, 2016.

86. Tess Russell, "Alone When the Bedbugs Bite," *New York Times*, November 18,
2010.

87. Schumer, *Girl with the Lower Back Tattoo*, 248.

88. Federal Bed Bug Workgroup, *Collaborative Strategy on Bed Bugs*.

89. Federal Bed Bug Workgroup, *Collaborative Strategy on Bed Bugs*.

90. Federal Bed Bug Workgroup, *Collaborative Strategy on Bed Bugs*.

91. Dini M. Miller, *Bed Bug Action Plan for Shelters* (Richmond: Virginia Department of Agriculture and Consumer Services, 2014), http://www.vdacs.virginia.gov/pdf
/bb-shelters1.pdf.

92. Sella, "Bedbugs in the Duvet."

93. Millard, quoted in Potter, "History of Bed Bug Management."

94. Javier C. Hernández, "In the War on Bedbugs, a New Strategy," *New York Times*,
July 28, 2010, https://cityroom.blogs.nytimes.com/2010/07/28/in-the-war-on-bedbugs
-a-new-attack-strategy/.

95. Lawrence Wright, *Warm and Snug: The History of the Bed* (London: Routledge &
Kegan Paul, 1962; Gloucestershire: Sutton, 2004).

96. Ralph Gardner, "Sniffing Out Tiny Terrorists," *Wall Street Journal*, January 28,
2011, https://www.wsj.com/articles/SB10001424052748704268104576108340156262716.

97. Jim Shea, "Bedbug Invasion Causes Panic," *Hartford Courier*, October 16, 2010,
http://articles.courant.com/2010-10-16/health/hc-shea-bedbugs-1016-20101016_1
_strains-africanized-potential-donors.

98. Goldstein, "An Army of Bedbugs."

99. "Bedbugs Are Vicious, Evil Little Creatures," *Tiny Frog: Atheism, Evolution,
Skepticism* (blog), September 2, 2008, https://tinyfrog.wordpress.com/2008/09/02
/bedbugs-are-vicious-evil-little-creatures/.

100. Quoted in Murphy, "Bedbugs Bad for Business?"

101. Saki Knafo, "The Man Who Lets Bedbugs Bite," *New York Times*, February 20,

2009, https://www.nytimes.com/2009/02/22/nyregion/thecity/22bedb.html; James Goddard quoted in Donald G. McNeil Jr., "They Crawl, They Bite, They Baffle Scientists, *New York Times*, August 30, 2020, https://www.nytimes.com/2010/08/31/science /31bedbug.html.

102. "Bed Bug and Beyond," *Daily Show with Jon Stewart*, September 18, 2015, http://www.cc.com/video-clips/5ay6pu/the-daily-show-with-jon-stewart-bed-bug ---beyond.

103. Bret Stephens quoted by Allen Smith, "A Professor Labeled Bret Stephens a 'Bedbug': Here's What the NYT Columnist Did Next," NBC News, accessed October 15, 2020, https://www.nbcnews.com/politics/politics-news/professor-labeled-bret -stephens-bedbug-here-s-what-nyt-columnist-n1046736.

104. Donald Trump quoted by Arren Kimbel-Sannit, "Trump Denies His Doral Resort Is Infested with Bed Bugs," Politico, August 27, 2019, https://www.politico .com/story/2019/08/27/president-trump-tweet-doral-florida-hotel-resort-bed-bugs -1475911.

105. Jane Brody, "The Pandemic May Spare Us from Another Plague: Bedbugs," *New York Times*, June 29, 2020, https://www.nytimes.com/2020/06/29/well/live /coronavirus-bedbugs.html.

Chapter 3 • Praying Lice

1. Plutarch, "The Life of L. C. Sylla," in *The Third Volume of Plutarch's Lives: Translated from the Greek, by Several Hands* (1693), 273.

2. Jan Bondeson, "Phthiriasis: The Riddle of the Lousy Disease," *Journal of the Royal Society of Medicine* 91 (1998): 328.

3. Nicholas Wade, "In Lice, Clues to Human Origins and Attire," *New York Times*, March 8, 2007.

4. Hans Zinsser, *Rats, Lice and History* (New York: Bantam Books, 1934, 1935), 134.

5. Amy Stewart, *Wicked Bugs: The Louse That Conquered Napoleon's Army and Other Diabolical Insects* (Chapel Hill, NC: Algonquin Books, 2011), 222–23.

6. Michelangelo Marisa da Caravaggio, *Martha and Mary Magdalene*, ca. 1598, oil and tempera on canvas, Detroit Institute of Art, https://www.dia.org/art/collection /object/martha-and-mary-magdalene-36204.

7. On the various pagan and Christians who died of the lousy disease, see J. R. Busvine, *Insects, Hygiene and History* (London: Athlone Press, 1976), 88–106.

8. Thomas Beard, *The Theatre of Gods Judgements wherein Is Represented the Admirable Justice of God against All Notorious Sinners* (1642), http://eebo.chadwyck. com.proxy.library.ucsb.edu:2048/search/fulltext?ACTION=ByID&ID=D00000 123256710000&SOURCE=var_spell.cfg&WARN=N&FILE=../session/1340220888 _6418.

9. Quoted in Busvine, *Insects, Hygiene and History*, 99.

10. Thomas Muffet, *The Theater of Insects*, vol. 3 of *The History of Four-Footed Beasts Serpents and Insects* (London: Printed by E. C., 1658; repr., New York: Da Capo Press, 1967), 3:1090.

11. Antoinette Bourignon, *The Light of the World a Most True Relation of a Pilgrimess Antonia Bourignon Travelling towards Eternity* (1696), 72.

12. Joseph Fletcher, *The Historie of the Perfect-Cursed-Blessed Man* (1628).

13. Thomas Hall, *Comarum Akosmia the Loathsomeness of Long Haire* (1654), 3–4, 47.

14. "A. B.," *Gentleman's Magazine* 16 (October 1746): 535.

15. "Further Observations on the Generation and Increase of the Said Little Animal," *Gentleman's Magazine* 17 (January 1747): 14.

16. Edward Cave, "To A. B.," *Gentleman's Magazine* 16 (December 1746): 660.

17. Peter Pindar, *The Lousiad: An Heroic-Comic Poem: Canto I* (Dublin: Colles, White, Byrne, W. Porter, Lewis, and Moore, 1786), 1, 8.

18. Pindar, *The Lousiad: Canto II* (London: T. Evans; Dublin: Robertson and Berry, 1793), 44–45.

19. Pinder, *Lousiad*, 45.

20. Nicholas Culpeper, *Culpeper's Directory for Midwives: Or, a Guide for Women. The Second Part* (1662), 133–34.

21. Antonie van Leeuwenhoek, *The Select Works of Antony Van Leeuwenhoek, Translated from the Dutch*, 2 vols. (London, 1800), 2:108–9.

22. Leeuwenhoek, "Of the Louse," in *Select Works*, 2:166.

23. Leeuwenhoek, "Of the Louse," in *Select Works*, 2:163.

24. Leeuwenhoek, "Of the Louse," in *Select Works*, 2:168.

25. Leeuwenhoek, "Of the Louse," in *Select Works*, 2:68–69.

26. On collecting and cabinets of curiosities, see Paula Findlen, *Possessing Nature* (Berkeley: University of California Press, 1994), and Lorraine Daston and Katharine Park, *Wonders and the Order of Nature, 1150–1750* (New York: Zone Books, 1998).

27. Théodore de Mayerne, "To the Noble Knight, and the King's Chief Physician, Dr. William Paddy," in Edward Topsell, *The History of Four-Footed Beasts and Serpents*, vol. 2 (London, 1658).

28. John Ray, *The Wisdom of God Manifested in the Works of Creation* (London: W. Innys, 1691), 309.

29. John Wilkins, *Of the Principles and Duties of Natural Religion Two Books* (London: T. Bassert, H, Brome, A. Chiswell, 1675).

30. Robert Hooke, *Micrographia: Some Physiological Descriptions of Minute Bodies Made with Magnifying Glasses with Observations and Inquiries Thereupon* (London: Jo. Martyn and Ja. Allestry, 1665), 193–94.

31. Hooke, *Micrographia*, 213.

32. Hooke, *Micrographia*, 211.

33. See Carolyn Merchant, *The Death of Nature* (New York: Harper and Row, 1980).

34. Hooke, "The Preface," in *Micrographia*.

35. Hooke, "The Preface," in *Micrographia*.

36. Thomas Sprat, *History of the Royal Society, for the Improving of Natural Knowledge (1667)*, ed. Jackson I. Cope and Harold Whitmore (St. Louis, MO: Washington University, 1958), 342–43.

37. Hooke, *Micrographia*, 213.

38. Andrew Marvell, "Instructions to a Painter about the Dutch Wars, 1667," in *The Poetical Works of Andrew Marvell* (London: Alexander Murray, 1870), https://books.google.com/books?id=LFMCAAAAQAAJ&dq=Marvell+Comptroller&source=gbs_navlinks_s.

39. Hooke, *Micrographia*, 146.

40. Margaret Cavendish, "Further Observations upon Experimental Philosophy," in *Observations upon Experimental Philosophy* (London: A. Maxwell, 1666), 12–13.

41. Cavendish, "Further Observations," 13.

42. Margaret Cavendish, *The Description of a New World, Called The Blazing World* (London: A. Maxwell,1666), 31.

43. Pietro Aretino, *The Wandring Whore*, vol. 2 (London, 1660), 7.

44. John Partridge, *The Widdowes Treasure* (London, 1595), http://eebo.chadwyck.com.proxy.library.ucsb.edu:2048/search/fulltext?SOURCE=var_spell.cfg&ACTION=ByID&ID=D00000331433960000&WARN=N&SIZE=117&FILE=../session/1385321812_3805&SEARCHSCREEN=CITATIONS&DISPLAY=AUTHOR&ECCO=N.

45. Vincenzo Gatti, "A Specimen of Miscellaneous Observations, on Medical Subjects," in *A Collection of Pieces Relative to Inoculation for the Small-Pox* (London, 1768), 201.

46. Daniel Heinsius, *Laus Pediculi, or an Apoloeticall Speech, Directed to the Worshipfull Masters and Wardens of Beggars Hall*, trans. James Guitard (London: Tho. Harper, 1634), 16.

47. Denise Grady, "Itching: More Than Skin Deep," *New York Times*, February 17, 2014, https://www.nytimes.com/2014/02/18/health/itching-more-than-skin-deep.html.

48. William Newcastle, "The Beggar's Marriage," in Margaret Cavendish, *Natures Pictures* (1654), 144–145. At this time, Newcastle may have been suffering from impotence due to syphilis, so this poem is particularly poignant. See Katie Whitaker, *Mad Madge: The Extraordinary Life of Margaret Cavendish, the First Woman to Live by Her Pen* (New York: Perseus Books, 2002), 100–101.

49. Charles Sackville, 6th Earl of Dorset, "The Duel of the Crabs," in *Poems on Affairs of State: Augustan Satirical Verse, 1660–1714*, ed. George deforest Lord (New Haven, CT: Yale University Press, 1963), 395.

50. Sackville, "Duel of the Crabs," 395.

Chapter 4 • Lousy Societies

1. Jonathan Swift, "The Story of the Injured Lady" (1746), Oxford Text Archive, https://ota.bodleian.ox.ac.uk/repository/xmlui/handle/20.500.12024/2780.

2. Robert Burns, "To a Louse" (1786), http://www.robertburnsfederation.com/poems/translations/552.htm.

3. In a book review in the *New York Times* (December 29, 2013) of Paul Bloom's *Little Angels: The Origins of Good and Evil* (New York: Crown, 2013), Simon Baron-Cohen writes, "Bloom explores the interesting overlap between feelings of disgust in relation to our food tastes and feelings of disgust in relation to our moral tastes, raising the intriguing idea that the same neural circuitry that allows disgust toward the former was co-opted in evolution to allow disgust toward the latter" (15). Hence we might argue that the disgust directed toward lice was redirected to those who carried lice— from nature to morality.

4. Thomas Muffet, *The Theater of Insects*, vol. 3 of *The History of Four-Footed Beasts Serpents and Insects* (London: Printed by E. C., 1658; repr., New York: Da Capo Press, 1967), 3:1102.

5. Keith Thomas, "Cleanliness and Godliness in Early Modern England," in *Religion, Culture and Society in Early Modern England*, ed. Anthony Fletcher and Peter Roberts (Cambridge: Cambridge University Press, 1994), 72.

6. "George Washington's Rules of Civility and Decent Behavior in Company and Conversation," *Foundations Magazine*, accessed March 22, 2015, http://www.founda tionsmag.com/civility.html.

7. Desiderius Erasmus, *The ciuilitie of childehode with the discipline and institucion of children, distributed in small and compe[n]dious chapiters*, trans. Thomas Paynell (1560), http://eebo.chadwyck.com.ezproxy.proxy.library.oregonstate.edu/search/full text?ACTION=ByID&ID=D20000230364410023&SOURCE=var_spell.cfg&DISPLAY =AUTHOR&WARN=N&FILE=../session/1585344170_15816. I have modernized the spelling in the quotation from this work.

8. Burns, "To a Louse."

9. On the meaning and history of hair, see Mary Douglas, *Purity and Danger: An Analysis of the Concepts of Pollution and Control* (London: Routledge & Kegan Paul, 1966); and Kurt Stenn, *Hair: A Human History* (New York: Pegasus Books, 2016).

10. Burns, "To a Louse."

11. Muffet, *Theater of Insects*, 3:1091.

12. Daniel Heinsius, *Laus Pediculis, or an Apoloeticall Speech, Directed to the Worshipfull Masters of Beggars Hall*, trans. James Guitard (1634).

13. Robert Heath, *Paradoxical Assertions and Philosophical Problems Full of Delight and Recreation for All Ladies and Youthful Fancies* (1659), 30–32.

14. Anonymous, *A Book to Help the Young and Gay, to Pass Tedious Hours Away* (London, 1750?).

15. John Hawkesworth, *The Adventurer*, vol. 2 (Dublin, 1754), 127–39.

16. Thomas Dekker, *The Belman of London Bringing to Light the Most Notorious Villanies That Are Now Practiced in the Kingdome* (London, 1608).

17. Edward Ward, *The Secret History of the London Clubs* (London, 1709), 229–31.

18. Lord Barrey, *Ram-Alley: Or Merrie-Trickes. A Comedy Divers Times Here-to-Fore Acted by the Children of the Kings Revels* (1611).

19. John Trusler, *The Works of William Hogarth: In a Series of Engravings with Descriptions, and a Comment on Their Moral Tendency* (London: Jones, 1833), 90, http://www.gutenberg.org/files/2500-h//2500-h.htm.

20. François Rabelais, *The Second Book of the Works of Mr Francis Rabelais, Doctor in Physick Treating of the Heroik Deeds and Sayings of the Good Pantagruel*, trans. S. T. U. C. (1653), 90.

21. Thomas Fuller, *The Appeal of Iniured Innocence* (1659), 2.

22. John Wilmot, Earl of Rochester, *The Irish Rogue* (1740), quoted in Lisa T. Sarasohn, "The Microscopist as Voyeur," in Sigrun Haude and Melinda S. Zook, eds., *Challenging Orthodoxies: The Social and Cultural Worlds of Early Modern Women* (Farnham, Surrey, UK: Ashgate, 2014), 18–19.

23. Victor Hugo, *Les Misérables*, trans. Julie Rose (New York: Modern Library, 2009), 746–48.

24. Emmanuel Le Roy Ladurie, *Montaillou: The Promised Land of Error*, trans. Barbara Bray (New York: Vintage Books, 1979), 141.

25. Margery Kempe, *The Book of Margery Kempe*, trans. Barry Windeatt (New York: Penguin Classics, 1986), 281. All of these discussions of medieval lice can be found in Virginia Smith, *Clean: A History of Personal Hygiene and Purity* (Oxford: Oxford University Press, 2007), 158–60.

26. Anonymous, *Aristotle's New Book of Problems*, 6th ed. (1725), 32.

27. Samuel Pepys, "Monday 18 July 1664," *Diary of Samuel Pepys*, https://www.pepysdiary.com/diary/1664/07/18/.

28. Janet Arnold, *Perukes and Periwigs: A Survey, c. 1660–1740* (London: Stationary Office Books, 1971), 5–7.

29. Samuel Pepys, "Saturday 2 May 1663, *Diary of Samuel Pepys*, https://www.pepysdiary.com/diary/1663/05/02/.

30. Matthew P. Davies, "The Tailors of London and Their Guild, c. 1300–1500," master's thesis (Corpus Christi College, University of Oxford, 1994).

31. Claire Tomalin, *Samuel Pepys: The Unequalled Self* (New York: Vintage Books, 2002), 117–18.

32. Samuel Pepys, "Sunday 8 February 1662/63," *Diary of Samuel Pepys*, https://www.pepysdiary.com/diary/1663/02/08/.

33. Samuel Pepys, "Saturday 23 January 1668/69," *Diary of Samuel Pepys*, https://www.pepysdiary.com/diary/1669/01/23./

34. William Andrews, *At the Sign of the Barber's Pole: Studies in Hirsute History* (Cottingham: J. R. Tutin, 1904), http://www.gutenberg.org/ebooks/19925.

35. Don Herzog, "The Trouble with Hairdressers," *Representations* 53 (1996): 23.

36. Lady Louisa Stuart, quoted in *Horace Walpole's Correspondence*, vol. 30, 275, Lewis Walpole Library, Yale University, http://images.library.yale.edu/hwcorrespondence/page.asp.

37. Samuel Pepys, "Saturday 6 June 1663," *Diary of Samuel Pepys*, https://www.pepysdiary.com/diary/1663/06/06/.

38. Jonathan Swift, *Gulliver's Travels* (Chicago: Children's Press, 1969), 131, 159.

39. Muffet, *Theater of Insects*, 3:1092.

40. Edmund Spenser, "A View of the State of Ireland (1596)," in Saint Edmund Campion, *Two Histories of Ireland* (1633), 38.

41. *Batman upon Bartholome His Booke De Proprietatibus Rerum* (1582), 231n.

42. William Mercer, *The Moderate Cavalier* (1675), 11. On the use of lice as a justification for decimation in Ireland and America, see Katie Kane, "Nits Make Lice," *Cultural Critique* 42 (1999): 81–103.

43. John Percival to Mr. Digby Cotes, September 18, 1701, quoted in *The English Travels of Sir John Percival and William Byrd II: The Percival Diary of 1701* (Columbus: University of Missouri Press, 1989), 187.

44. On attitudes toward *plica polonica*, see Eglé Sakalauskaité-Juodeikiené, "*Plica Polonica* through the Centuries the Most 'Horrible, Incurable, and Unsightly,'" *World Neurology*, March 25, 2020, https://worldneurologyonline.com/article/plica-polonica-through-the-centuries-the-most-horrible-incurable-and-unsightly/; "Nontoxic Head Lice Treatment," Nuvo Method for Head Lice, revised January 3, 2008, https://nuvoforheadlice.com/?page_id=120.

45. Pierre Chevalier, *A Discourse of the Original, Countrey, Manners, Government and Religion of the Cossacks with Another of the Precopian Tartars* (1672), 24.

46. R. W. Gwadz, "Parasitology 8, Arthropods of Medical Importance," transcribed by Ian Cohen, Nuvo Method for Head Lice, accessed November 10, 2020, https://nuvoforheadlice.com/?page_id=122.2003.

47. Andrew Duncan, *Annals of Medicine for the Year 1796* (Edinburgh, 1799), 7–8.

48. Michael Adams, *The New Royal Geographical Magazine* (1794), 193.

49. *Memoirs of the Royal Society; Being a New Abridgment of the Philosophical Transactions*, 10 vols. (London, 1738/39), 9:160–61.

50. Thomas Hall, "To the Long-hair'd Gallants of These Times," in *Comarom Akosmia, The Loathsomnesse of Long Hair* (1654), https://quod.lib.umich.edu/e/eebo /A45331.0001.001?view=toc.

51. F. L. Fontaine, "Surgical and Medical Treatises on Various Subjects for the Year 1796," in *Annals of Medicine*, vol. 1 (London, 1799), 13.

52. Pietro Matire d'Anghiera, *The Decades of the New World or West Indies*, trans. Richard Eden, in John Andrewes, *A Golden Trumpet Sounding an Alarm to Judgement* (1648), 186.

53. Gonzalo Fernando de Oviedo, *Voyages and Travels to the New World . . . , The Fifth Book* in *Purchas His Pilgrimes, The First Book* (London, 1625), 975.

54. Samuel Hearne, *A Journey from Prince of Wale's Fort in Hudson's Bay, to the Northern Ocean* (London, 1795), 325–26.

55. Fernando de Oviedo, *Voyages and Travels*, 975.

56. Miguel de Cervantes, *The History of the Valorous and Witty Knight-Errant, Don Quixote, of the Mancha Tr. Out of the Spanish* (1652).

57. Peter Kolben, *The Present State of the Cape of Good Hope or a Particular Account of the Several Nations of the Hottentots*, vol. 1, trans. Mr. Medley (1731), 203–4.

58. On the "truths" of Kolben's narrative, see Damien Shaw, "A Fraudulent Truth? Christian Damberger's Vision of Africa (1801)," *English Studies in Africa* 60, no. 2 (2017): 1–11.

Chapter 5 • The Perils of Lice in the Modern World

1. Charles Darwin, "Island on Chiloe, July 1834," in *Charles Darwin's Zoology Notes and Specimen Lists from H.M.S. Beagle* (Cambridge: Cambridge University Press, 2004), 283.

2. Charles Darwin, "Pediculus. Chiloe. July," Darwin Online, accessed May 2, 2020, darwin-online.org.uk/content/frameset?viewtype=side&itemID=CUL-DAR29.1.C2 &pageseq=1.

3. For the history of science and racial theories, see John P. Jackson and Nadine M. Weldman, *Race, Racism, and Science* (Santa Barbara, CA: ABC-CLIO, 2004).

4. Pluralism and polygenesis are discussed at length in Adrian Desmond and James Moore, *Darwin's Sacred Cause: How a Hatred of Slavery Shaped Darwin's Views on Human Evolution* (Boston: Houghton Mifflin, 2009).

5. *Punch, or the London Charivari* (London, 1852).

6. A. S. Packard Jr., "Certain Parasitic Insects," *American Naturalist* 4 (1871): 67.

7. Charles Darwin, *The Descent of Man* (1871), https://ia600302.us.archive.org/10 /items/thedescentofman02300gut/dscmn10.txt.

8. P. N. K. Schwenk, "Phthiriasis, with Report of Cases of Phthiriasis Pubis in Eye-lashes, Eye-brows and Head," *Times and Register* 22 (May 9, 1891): 381–83.

9. Monroe Woolley, "Style—Its Follies and Cost: A Review of the Present Over-Civilized Order of Things in the Matter of Dress," *Health* (1900–1913: July 1912), 62, 7; *American Periodicals*, 180.

10. On Nicolle, see Kim Pellis, *Charles Nicolle, Pasteur's Imperial Missionary: Typhus and Tunisia* (Rochester, NY: Rochester University Press, 2006). The bacterium

that causes typhus, *Rickettsia prowazeki*, was detected independently in 1916 by the Brazilian pathologist and physician Henrique da Rocha Lima, who named it after two early typhus researchers, the American scientist Henry Ricketts and the Czech bacteriologist Stanislaw von Prowazek. Both achieved immortality in the name of the disease, although their immediate reward was to die of its ravages.

11. The classic account of the history of typhus is Hans Zinsser, *Rats, Lice and History* (New York: Little, Brown, 1935; repr., New York: Black Dog and Leventhal, 1963). (All quotations in this chapter come from the 1963 edition.) More recent accounts can be found in Alex Bein, *The Jewish Parasite: Notes on the Semantics of the Jewish Problem, with Special Reference to Germany* (New York: Leo Baeck Institute, 1964), 3–39; Sir Richard Evans, "The Great Unwashed," Gresham College, February 26, 2013, http://www.gresham.ac.uk/lectures-and-events/the-great-unwashed; and Amy Stewart, *Wicked Bugs: The Louse That Conquered Napoleon's Army and Other Diabolical Insects* (Chapel Hill, NC: Algonquin Books, 2011), 222–27.

12. Evans, "The Great Unwashed."

13. P. Brouqui and D. Raoult, "Arthropod-borne Diseases in the Homeless," *Annals of the New York Academy of Sciences* 10, no. 1078 (2006): 223–35.

14. Paul Fussell, in *The Great War in Modern Memory* (London: Oxford University Press, 1975), describes World War I in particular as an ironic war (8).

15. Zinsser, *Rats, Lice and History*, 183–84.

16. Zinsser, *Rats, Lice and History*, 188.

17. Edward Long, *The History of Jamaica or, General Survey of the Antient and Modern State of the Island* (1774), 382. For a more detailed discussion of Long and other American naturalists, see John C. Greene, "The American Debate on the Negro's Place in Nature," *Journal of the History of Ideas* 15 (1954): 384–96.

18. Charles White, *An Account of the Gradual Gradation in Man, and in Different Animals and Vegetables* (1799), 79.

19. Charles Darwin, *The Descent of Man, and Selection in Relation to Sex*, 2nd ed. (New York: D. Appleton, 1889), 170.

20. Desmond and Moore, *Darwin's Sacred Cause*, 193.

21. Charles Darwin to Henry Denny, January 17, 1865, Darwin Correspondence Project, University of Cambridge, https://www.darwinproject.ac.uk/letter/?docId =letters/DCP-LETT-4747F.xml;query=Henry%20Denny;brand=default.

22. Darwin, *Descent of Man*, 167.

23. Henry Denny to Charles Darwin, January 23, 1865, Darwin Correspondence Project, University of Cambridge, https://www.darwinproject.ac.uk/letter/?docId =letters/DCP-LETT-4753.xml;query=Henry%20Denny;brand=default.

24. Henry Denny, *Monographia Anoplurorum Britanniae, or An Essay on the British Species of Parasitic Insects* (London: H. G. Bohn, 1842), 18.

25. A. S. Packard, "Certain Parasitic Insects," *American Naturalist* 4, no. 2 (1870): 83–99, http://www.jstor.org/stable/2446722.

26. Samuel Clemens to Jane Lampton Clemens, March 20, 1862, Mark Twain Project, http://www.marktwainproject.org/xtf/view?docId=letters/UCCL00040. xml;query=;searchAll=;sectionType1=;sectionType2=;sectionType3=;sectionType4= ;sectionType5=;style=letter;brand=mtp#1.

27. Gordon Floyd Ferris, *Contributions towards a Monograph on the Sucking Lice: Part 1* (Stanford, CA: Stanford University Press, 1919); Gordon Floyd Ferris, *Contributions towards a Monograph on the Sucking Lice: Part 8* (Stanford, CA: Stanford University Press, 1935), 567–78. I wish to thank Dr. Richard Pollack for this reference and for his "welcome to the confusing and maddening world of louse systematics" (personal correspondence, October 29, 2013). Most modern entomologists think that lice come in all shades, from beige to black on all types of bodies.

28. Heinrich Himmler, quoted in Hugh Raffles, "Jews, Lice, and History," *Public Culture* 19, no. 3 (2007): 521.

29. Raffles, "Jews, Lice, and History," 522.

30. Alexander Cockburn, "Zyklon B on the U.S. Border," *The Nation*, June 21, 2007, https://www.thenation.com/article/zyklon-b-us-border/.

31. Nanette Blitz, quoted in Stephanie Pappas, "Anne Frank Likely Died Earlier Than Believed," *Live Science*, April 2, 2015, https://www.livescience.com/50360-anne -frank-died-earlier.html.

32. D. Raout, "Outbreak of Epidemic Typhus Associated with Trench Fever in Burundi," *Lancet* 352, no. 9125 (August 1, 1998): https://www.ncbi.nlm.nih.gov/pub med/9717922.

33. *Modern Family*, episode 111, "The Feud," written by Christopher Lloyd and Dan O'Shannon, aired February 26, 2014, on ABC.

34. Isaac Rosenberg, "The Louse-Hunting," Poetry Foundation, accessed October 10, 2015, https://www.poetryfoundation.org/poems/47413/louse-hunting.

35. Siegfried Sassoon, "Suicide in the Trenches," Poem Hunter, accessed November 5, 2020, https://www.poemhunter.com/poem/suicide-in-the-trenches/.

36. Helen Zenna Smith, *Not So Quiet . . . Stepdaughters of War* (1930; repr., New York: Feminist Press, 1989), 239.

37. Smith, *Not So Quiet*, 17. See the afterword, by Jane Marcus, for an account of the gender unorthodoxies in the novel (241–93).

38. Erich Maria Remarque, *All Quiet on the Western Front*, trans. W. Wheen Fawcett Crest (New York: Little, Brown, 1929), https://docs.google.com/viewer?a=v&pid=sites &srcid=YXBwcy5kaXN0cmljdDgzMy5vcmd8Y2FyYm9uZWVuZ2xpc2h8Z3g6NzMz YWI5MGM0NDNiN.

39. "Cure for Lice in the Trenches," *Daily Mirror*, May 9, 1916, http://skittishlibrary .co.uk/cure-for-lice-in-the-trenches-1915/.

40. *Daily Mirror*, May 9, 1916.

41. Julia Nurse, "A Commemoration of Armistice Day," Wellcome Library, November 11, 2014, http://blog.wellcomelibrary.org/2014/11/a-commemoration-of-armistice -day/.

42. A. E. Shipley, "Insects and War: Lice," *British Medical Journal* 2, no. 2803 (September 19, 1914): 497–99, https://www.jstor.org/stable/25311196?seq=1#page_scan _tab_contents.

43. *Under the Rainbow: A History of Its Service in the War against Germany*, Battery F, 150th F. A. (Indianapolis: Hollenbeck Press, 1919), 135–36, https://books.google.com /books?id=Pa2fAAAAMAAJ&printsec=frontcover&source=gbs_ge_summary_r&cad =0#v=onepage&q=cootie&f=false.

44. A. A. Milne, quoted in "A. A. Milne in the Great War," Science Fiction and Fantasy Writers in the Great War, accessed February 8, 2018, https://fantastic-writers -and-the-great-war.com/war-experiences/a-a-milne/.

45. Arthur Allen, *The Fantastic Laboratory of Dr. Weigl: How Two Brave Scientists Battled Typhus and Sabotaged the Nazis* (New York: W. W. Norton, 2014), 25–29; Zinsser, *Rats, Lice and History*, 296–301.

46. Zinsser, *Rats, Lice and History*, 172.

47. Zinsser, *Rats, Lice and* History, 125.

48. Hans Zinsser, *Rats, Lice and History, with a New Introduction by Gerald Grob* (New Brunswick, NJ: Transaction Publishers, 2008), xiv.

49. Vladimir Lenin, quoted in Frederick Holmes, "Medicine in the First World War: Typhus on the Eastern Front," University of Kansas Medical Center, accessed February 8, 2019, http://www.kumc.edu/wwi/index-of-essays/typhus-on-the-eastern-front .html.

50. See Albert Rhys Williams, *Through the Russian Revolution* (London: Boni and Liveright, 1921), 282, for a copy of the poster.

51. The image is duplicated in Evans, "The Great Unwashed."

52. Allen, *Fantastic Laboratory*, 26.

53. Allen, *Fantastic Laboratory*, 35.

54. Winston Churchill, quoted in Allen, *Fantastic Laboratory*, 35.

55. Andrew Dewar Gibb, *With Winston Churchill at the Front: Winston on the Western Front 1916* (Barnsley, UK: Frontline Books, 2016), https://books.google.com /books?id=aSZGDAAAQBAJ&pg=PT26&dq=Andrew+Dewar+Gibb&source=gbs_toc _r&cad=4#v=onepage&q=lice&f=false.

56. Boris Johnson, "The Woman Who Made Winston Churchill," *The Telegraph*, October 12, 2014, https://www.telegraph.co.uk/news/politics/conservative/11155850 /Boris-Johnson-the-woman-who-made-Winston-Churchill.html.

57. Howard Markel and Alexandra Minna Stern, "The Foreignness of Germs: The Persistent Association of Immigrants and Disease in American Society," *Milbank Quarterly* 80, no. 4 (2002): 757–83.

58. *New York Times*, July 17, 1921, quoted in Marcus Doel, *Geographies of Violence: Killing Space, Killing Time* (Thousand Oaks, CA: Sage, 2017), 121.

59. Quoted in Howard Markel, *Quarantine! Eastern European Jewish Immigrants and the New York City Epidemic of 1892* (Baltimore: Johns Hopkins University Press, 1999), 50.

60. *Washington Times*, May 4, 1919.

61. Markel and Stern, "Foreignness of Germs," 761.

62. John Burnett, "The Bath Riots: Indignity along the Mexican Border," *NPR*, January 28, 2006, https://www.npr.org/templates/story/story.php?storyId=5176177.

63. "Cecile Klein-Pollack Describes Arrival at Auschwitz," *Holocaust Encyclopedia*, US Holocaust Memorial Museum, accessed February 9, 2019, https://encyclopedia .ushmm.org/content/en/oral-history/cecilie-klein-pollack-describes-arrival-at -auschwitz.

64. Paul Julian Weindling, *Epidemics and Genocide in Eastern Europe, 1890–1945* (Oxford: Oxford University Press, 2000), 312.

65. Weindling, *Epidemics and Genocide*, 315.

66. Primo Levi, *The Complete Works of Primo Levi*, vol. 1 (London: Liveright, 2015), 238.

67. Weindling, *Epidemics and Genocide*, 305.

68. Friedrich Paul Berg, "The German Delousing Chambers," Institute for Historical Review, accessed November 8, 2016, http://www.ihr.org/jhr/v07/v07p-73_Berg.html.

69. Charles Nicolle, quoted in Pellis, *Charles Nicolle, Pasteur's Imperial Missionary*, 191.

70. Robert White-Stevens, quoted in "Rachel Carson's Lethal Claptrap," The Atheist Conservative, May 28, 2014, https://theatheistconservative.com/2014/05/28/rachel-carsons-lethal-claptrap/.

71. Steven Hayward, quoted in "Rachel Carson's Lethal Claptrap."

72. Quoted in Michael Savage, *Boosting Your Immunity against Infectious Diseases from the Flu and Measles to Tuberculosis* (New York: Hachette, 2016).

73. Todd Starnes, "Medical Staff Warned: Keep Your Mouths Shut about Illegal Immigrants or Face Arrest," Fox News, July 2, 2014, https://www.foxnews.com/opinion/medical-staff-warned-keep-your-mouths-shut-about-illegal-immigrants-or-face-arrest.

74. Janis Hootman, "Don't Mix Metaphors," *AAP News*, May 1, 1997, http://www.aappublications.org/content/13/5/32.1.

75. "Nit-Picking a Lousy Policy," *New York Times*, May 14, 2014, https://www.nytimes.com/roomfordebate/2014/05/14/nit-picking-a-lousy-policy.

76. "Nit-Picking a Lousy Policy."

77. "Nit-Picking a Lousy Policy."

78. AirAllé website, accessed November 30, 2016, https://airalle.com/airalle/.

79. Barry R. Pittendrigh, John M. Clark, J. Spencer Johnston, Si Hyeock Lee, Jeanne Romero-Severson, and Gregory A. Dasch, "Proposal for the Sequencing of a New Target Genome: White Paper for a Human Body Louse Genome Project," University of Geneva, Zdobnov's Computational Evolutionary Genomics Group, accessed November 10, 2020, http://cegg.unige.ch/system/files/BodyLouseGenomeWhitePaper.pdf.

80. Insect Research and Development Ltd. website, accessed November 30, 2016, http://insectresearch.com/.

81. Identify Us website, accessed November 30, 2016, https://identify.us.com/.

82. "Brazilian Waxes Could Make Pubic Lice Go Extinct," Smithsonian, January 14, 2013, https://www.smithsonianmag.com/smart-news/brazilian-waxes-could-make-pubic-lice-go-extinct-1258409/.

83. J. R. Busvine, *Insects, Hygiene and History* (London: Athlone Press, 1976), 194–95.

84. Franz Kafka, *The Metamorphosis: A New Translation by Susan Bernofsky* (New York: W. W. Norton, 2014), 121.

Chapter 6 • The Flea in Humanity's Ear

1. James Boswell, *Life of Samuel Johnson*, ed. Charles Grosvenor Osgood (1791; New York: Firework Press, 2015), 302.

2. May R. Berenbaum, *Ninety-nine Gnats, Nits, and Nibblers* (Urbana: University of Illinois Press, 1990), 216–17; and "Flea-Flickers and Football Fields," *American Entomologist* 54, no. 3 (2008): 132–33.

3. Thomas Muffet, *The Theater of Insects*, vol. 3 of *The History of Four-Footed Beasts Serpents and Insects* (London: Printed by E. C., 1658; repr., New York: Da Capo Press, 1967), 3:1101–3.

4. Muffet, *Theater of Insects*, 3:1102.

5. James Murray et al., *A New Historical Dictionary Founded on Historical Principles* (Oxford: Oxford University Press, 1901), 4:306.

6. Anonymous, "Upon the Biting of a Flea (c. 1650), quoted in Todd Andrew Borlik, ed. *Literature and Nature in the English Renaissance: An Ecocritical Anthology* (Cambridge: Cambridge University Press, 2019), 175–76.

7. Berenice Williams, "One Jump ahead of the Flea," *New Scientist*, July 31, 1966, 37.

8. Aesop, *The Flea and the Man*, ed. Charles Grosvenor Osgood (New York: Doubleday, Page, 1916), 35.

9. Jonathan Swift, "On Poetry: A Rhapsody," The Literature Network, accessed March 31, 2015, http://www.online-literature.com/swift/3515/.

10. Michel de Montaigne, iZ Quotes, accessed April 1, 2015, http://izquotes.com/quote/348141.

11. Aristophanes, "The Clouds" (419 BCE), Internet Classics Library, http://classics.mit.edu/Aristophanes/clouds.html.

12. Desiderius Erasmus, *The Praise of Folly*, trans. John Wilson (1668).

13. Erasmus, quoted in Brendan Lehane, *The Compleat Flea* (New York: Viking Press, 1969), 19.

14. Quoted in Lehane, *The Compleat Flea*, 19.

15. Peter Woodhouse, *The Flea Sic Parua Compenere Magnis*, Early English Books Online (London: John Smethwick, 1605), https://search-proquest-com.ezproxy.proxy.library.oregonstate.edu/eebo/docview/2240929539/fulltextPDF/37280AF3A34B4FA0PQ/3?accountid=13013.

16. Woodhouse, *The Flea*.

17. Woodhouse, *The Flea*.

18. "The Flea," *Harper's New Monthly Magazine* 19 (1859): 178.

19. Thomas Amory, *The Life of John Buncle: Esq.: Containing Various Observations and Reflections, Made in Several Parts of the World; and Many Extraordinary Reflections* (1755; repr., London: George Routledge and Sons, 1904), 152–53, https://quod.lib.umich.edu/e/ecco/004870786.0001.000?view=toc.

20. Muffet, *Theater of Insects*, 3:1101.

21. Robert Burton, *Versatile Ingenium: The Wittie Companion* (1679), 2.

22. Keith Moore, "The Ghost of a Flea," *The Repository* (blog), October 18, 2012, http://blogs.royalsociety.org/history-of-science/2012/10/18/ghost-of-a-flea/.

23. Brian W. Ogilvie, "Attending to Insects: Francis Willughby and John Ray," *Notes and Records of the Royal Society Journal of the History of Science* 66 (2012): 366, doi:10.1098/rsnr.2012.0051.

24. Anonymous, *Memoirs and Adventures of a Flea*, vol. 2 (London, 1785), 36.

25. Anonymous, *A Book to Help the Young and Gay, to Pass the Tedious Hours Away* (London, 1750?), 119.

26. Peter Pindar, "An Elegy to the Fleas of Tenreriffe," in *The Works of Peter Pindar in Two Volumes* (Dublin, 1792), 2:369.

27. Muffet, "Preface," in *Theater of Insects*, 2Av.

28. Antony Van Leeuwenhoek, "Of the Flea," in *The Select Works of Antony Van Leeuwenhoek*, ed. Samuel Hoole (London: Philanthropic Society, 1808), 43.

29. Van Leeuwenhoek, "Of the Flea," 33–46.

30. Robert Hooke, "Preface," in *Micrographia* (1665), http://www.gutenberg.org /files/15491/15491-h/15491-h.htm.

31. Robert Hooke, "Of a Flea," in *Micrographia*.

32. Hooke, "Of a Flea," in *Micrographia*.

33. Moore, "Ghost of a Flea."

34. Quoted in G. E. Bentley Jr., *Blake Records* (Oxford: Clarendon Press, 1969), 39.

35. William Blake, quote in Alexander Gilchrist and Anne Gilchrist, *Life of William Blake*, 2 vols. (London: Macmillan, 1880), https://en.wikisource.org/wiki/Life_of _William_Blake_(1880),_Volume_1/Chapters_28%E2%80%9430.

36. G. K. Chesterton, *William Blake* (London: Duckworth, 1910), 154.

37. In *William Blake: The Critical Heritage*, ed. Gerald Eades Bentley (London: Routledge, 1975–95), 168–69.

38. Thomas Shadwell, *The Virtuoso*, ed. Marjorie Hope Nicolson and David Stuart Rodes (Lincoln: University of Nebraska Press, 1966), 31.

39. Reverend Lynam, ed., *The British Essayists*, vols. 1–3 (London: J. F. Dove, 1827), III:234.

40. "Extract of a Letter from a Gentleman in Maryland, to His Friend in London," in Granville Sharp, *The Just Limitations of Slavery: In the Laws of God, Compared with the Unbounded Claims of the African Traders and British American Slaveholders* (London: B. White, 1776), 43.

41. Thomas Atwood, *History of Domenica* (1791), Digitizing Sponsor: Brown University, https://archive.org/details/historyofislando00atwo.

42. Francisco de Oviedo, quoted in Amy Stewart, *Wicked Bugs: The Louse That Conquered Napoleon's Army and Other Diabolical Insects* (Chapel Hill, NC: Algonquin Books, 2011), 77.

43. Stewart, *Wicked Bugs*, 78.

44. Muffet, *Theater of Insects*, 3:1102.

45. On "flea-lore," see H. David Brumble, "John Donne's 'The Flea': Some Implications of the Encyclopedic and Poetic Flea Traditions," *Critical Quarterly* 15 (1973): 147–54.

46. John Donne, "The Flea," in *Poems of John Donne*, vol. 1, ed. E. K. Chambers (London: Lawrence and Bullen, 1896), 1.

47. John Donne, "A Defense of Womens Inconstancy," in *Juvenilia: Or Certain Paradoxes and Problems* (1633), https://quod.lib.umich.edu/e/eebo/A36301.0001.001 /1:4.1?rgn=div2;view=fulltext.

48. Quoted in Brumble, "John Donne's 'The Flea,' " 148.

49. Christopher Marlowe, *The Tragical History of Doctor Faustus from the Quarto of 1616*, ed. Alexander Dyce (1616), http://www.gutenberg.org/files/811/811-h/811-h.htm.

50. Muffet, *Theater of Insects*, 3:1101.

51. Etienne Pasquier, *La Puce de Madame de Roche*, quoted in Ann Rosalind Jones, "Contentious Readings: Urban Humanism and Gender Differences in *La Puce de Madame Des-Roches*," *Renaissance Quarterly* 48 (1995): 123.

52. Catherine des Roche, "Epitaph 1," in *From Mother and Daughter: Poems,*

Dialogues and Letters of Les Dames des Roche, ed. and trans. Anne R. Larsen (Chicago: University of Chicago Press, 2006), 178–79.

53. Editor Anne Larsen analyzes the political as well as the erotic meanings of Catherine des Roche's poem in *From Mother and Daughter*, 136–38.

54. John Donne the Younger, "Device," quoted in *Fleas in Amber: Verses and One Fable in Prose on the Philosophy of Vermin* (London: Fanfrolico Press, 1933).

55. William Cavendish, "The Varietie," in *The Country Captaine and the Varietie* (1649), 28.

56. Crissy Bergeron, "Georges de la Tour's *Flea-Catcher* and the Iconography of the Flea-Hunt in Seventeenth-Century Baroque Art," master of arts thesis (Louisiana State University, 2007), 32–43.

57. *Memoirs and Adventures of a Flea*, 25.

58. *Memoirs and Adventures of a Flea*, 51.

59. *Memoirs and Adventures of a Flea*, 50–51.

60. Anonymous, "Upon the Biting of a Flea" (ca. 1650), quoted in Borlik, *Literature and Nature in the English Renaissance*, 176.

Chapter 7 • *Fleas Become Killer Comedians*

1. Baronne d'Oberkirche, quoted in Caroline Walker, *Queen of Fashion: What Marie Antoinette Wore to the Revolution* (New York: Henry Holt, 2007), 117. *Puce* is French for flea.

2. Walker, *Queen of Fashion*, 256.

3. "The Flea," *Harper's New Monthly Magazine* 19 (1859): 178–80.

4. James Roberts, *The Narrative of James Roberts* (Chicago, 1858), www.deepsouth .unc.edu/neh/roberts/roberts.html.

5. *The Pacific Unitarian*, vols. 23–24 (1914), 71, https://play.google.com/books /reader?id=-YEfAAAAYAAJ&hl=en&pg=GBS.PA71.

6. Quoted in Ida Tarbell, "The American Woman," *American Magazine* 69 (1909): 475.

7. Harold Russell, *The Flea* (London: H. K. Lewis, 1913), 72.

8. "U.S. Dog Owners Fear Arrival of Africanized Fleas," *The Onion*, March 23, 2005, https://www.theonion.com/u-s-dog-owners-fear-arrival-of-africanized-fleas -1819567797.

9. "We Want a Union, Heckler's Fleas," *New Yorker* 63 (1946).

10. Quoted on the website Sideshow World, accessed April 6, 2019, http://www .sideshowworld.com/46-Flea%20Circus/2014/Heckler/Flea-Circus.html. For more on Heckler's flea circus, see "Trained Fleas Now Showing at Hubert's Museum," Duke Medical Center Archives, https://archives.mc.duke.edu/blog/it-came-archives-trained -fleas-now-showing-hubert%E2%80%99s-museum, and "The Flea Circus," Sideshow World, http://www.sideshowworld.com/46-Flea%20Circus/3-Prof-W-Hecklers /FleaCircus-Prof-W-Hecklers.html.

11. David J. Bibel and T. H. Chen, "Diagnosis of Plague: An Analysis of the Yersin-Kitasato Controversy," *Bacteriological Review* (1976): 633–51; Miriam Rothschild, *Dear Lord Rothschild: Birds, Butterflies and History* (Philadelphia: Balaban, 1983), 170–73; Kristin Johnson, *Ordering Life: Karl Jordan and the Naturalist Tradition* (Baltimore: Johns Hopkins University Press, 2005).

12. Jeffrey A. Lockwood, *Six-Legged Soldiers: Using Insects as Weapons of War* (Oxford: Oxford University Press, 2009), 108–27, 177–86.

13. Boris V. Schmid and Ulf Bütgen, "Climate-Driven Introduction of the Black Death and Successive Plague Reintroductions into Europe," *Proceedings of the National Academy of Sciences of the United States of America* 112, no. 10 (2015): doi:10.1073/pnas .1412887112.

14. Frank Moore, ed., *The Civil War in Song and Story, 1860–1865* (New York: Peter Fenelon Collier, 1865), 409.

15. "The Flea," 180, http://books.google.com/books/about/Harper_s_Magazine .html?id=m3kCAAAAIAAJ.

16. Henry Ward Beecher, *Treasury of Thought: Forming an Encyclopedia of Quotations from Ancient and Modern Authors*, 10th ed., ed. Maturin M. Ballou (Boston: Houghton Mifflin, 1881), 478.

17. For information about Louis Bertolotto and other nineteenth-century flea circuses, see "Historical Flea Circuses," Flea Circus Research Library, http://www .fleacircus.co.uk/History.htm.

18. Louis Bertolotto, *The History of the Flea, Notes, Observations and Amusing Anecdotes*, 2nd ed. (London: Crozier, 1835).

19. Francis T. Buckland, *Curiosities of Natural History: Fourth Series* (New York: Cosimo Classics, 1888), 115.

20. Advertisement for Louis Bertolotto's "Industrious Flea Circus," Flea Circus Research Library, accessed November 10, 2020, http://www.fleacircus.co.uk/History Bertolotto.htm.

21. Charles Dickens, *The Mudfog and Other Sketches* (1903), http://www.gutenberg .org/files/912/912-h/912-h.htm.

22. Dickens, *Mudfog and Other Sketches*.

23. Anonymous, *The Autobiography of a Flea* (1887), http://en.wikisource.org/wiki /The_Autobiography_of_a_Flea.

24. Johann Wolfgang Goethe, *Faust*, in *The Works of Goethe*, vol. 1 (1902), 102, https://books.google.com/books?id=gd8MAQAAIAAJ.

25. On the political elements in *Master Flea*, see Val Scullion and Marion Treby, "Repressive Politics and Satire in E. T. A. Hoffmann's 'Little Zaches' and 'Master Flea,'" *Journal of Law and Politics* 6 (2013): 133–45.

26. E. T. A. Hoffmann, "Second Adventure," in *Master Flea* (London: Thomas Davison, 1826), http://www.gutenberg.org/files/32223/32223-h/32223-h.htm.

27. Hans Christian Andersen, "The Flea and the Professor," *Scribner's Monthly* 6 (1873): 759–61.

28. Bertolotto, *History of the Flea*.

29. David Watson, "The Flea, the Catapult, and the Bow," FT Exploring Science and Technology, accessed February 12, 2019, http://ftexploring.com/-ftexplor/lifetech /flsbws1.html.

30. Walt Noon, Flea-Circus.com, https://www.noonco.com/flea/.

31. Walt Noon's flea circus can be viewed at Flea-Circus.com.

32. "Thundering Fleas," Internet Archive video, uploaded March 8, 2012, https:// archive.org/details/ThunderingFleas.

33. Stan Laurel and Oliver Hardy, "The Chimp," YouTube video, October 13, 2018, https://www.youtube.com/watch?v=tm20zVG4d7c.

34. Fred Allen, "It's in the Bag," YouTube video, December 31, 2010, https://www.youtube.com/watch?v=6sOLCUPj1-Q.

35. See the Jurassic Outpost website, accessed February 12, 2019, http://jurassic outpost.com/wp-content/uploads/2016/05/JurassicPark-Final.pdf.

36. Jim Frank, quoted in Ernest B. Furgurson, "A Speck of Showmanship: Is That *Pulex irritans* Pulling the Carriage, or Is It Someone Just Pulling Our Leg," *American Scholar*, June 3, 2011, https://theamericanscholar.org/a-speck-of-showmanship/#.

37. Adam Gertsacon, quoted in Jennifer Billock, "Revive the Charm of an 1800s Show with These Modern-Day Flea Circuses," *Smithsonian Magazine*, November 29, 2017, https://www.smithsonianmag.com/travel/modern-day-flea-circuses-180967355/.

38. Maria Franada Cardoso, "Museum of Copulatory Organs," YouTube video, July 19, 2012, https://www.youtube.com/watch?v=6J1srEtZr-I.

39. Melanie Kembrey, "The Amazing Maria Fernanda Cardoso," *Sydney Morning Herald*, November 16, 2018, https://www.smh.com.au/entertainment/art-and-design /the-amazing-maria-fernanda-cardoso-20181112-p50fhc.html.

40. Bertolotto, quoted by Richard Wiseman, "Staging a Flea Circus," NanoPDF.com, April 21, 2018, https://nanopdf.com/download/complete-history-of-the-circus_pdf.

41. Miriam Rothschild and Theresa May, *Fleas, Flukes and Cuckoos: A Study of Bird Parasites* (New York: Macmillan, 1957), 77.

42. Johnson, *Ordering Life*, 248.

43. "Siphonaptera Collection," Natural History Museum, accessed February 10, 2019, http://www.nhm.ac.uk/our-science/collections/entomology-collections/siphon aptera-collections.html.

44. Miriam Rothschild, "Nathaniel Charles Rothschild, 1877–1923," in *Fleas*, ed. R. Traub and H. Starcke (Rotterdam: A. A. Balkema, 1980), 1–3.

45. Miriam Rothschild, *Dear Lord Rothschild: Birds, Butterflies and History* (Glenside, PA: Balaban, 1983), 102–9.

46. Charles Rothschild, quoted in Johnson, *Ordering Life*, 42.

47. Rothschild and May, *Fleas, Flukes and Cuckoos*, 1–2.

48. Adolph Hitler, quoted in Richard Koenigsberg, "Hitler, Lenin—and the Desire to Destroy 'Parasites,'" Library of Social Science, March 6, 2015, https://www.library ofsocialscience.com/newsletter/posts/2015/2015-03-06-hitler-lenin.html.

49. "The Jew as World Parasite," German Propaganda Archive, https://research .calvin.edu/german-propaganda-archive/weltparasit.htm.

50. Rothschild and May, *Fleas, Flukes and Cuckoos*, 4.

51. L. Fabian Hirst, "Plague Fleas, with Special Reference to the Milroy Lectures, 1924," *Epidemiology and Infection* 24, no. 1 (1924): 1.

52. Charles Rothschild (1901), quoted in Rothschild, *Dear Lord Rothschild*, 171.

53. Charles Rothschild (1901), quoted in Rothschild, *Dear Lord Rothschild*, 171.

54. H. Maywell-Lefroy, *Indian Insect Life: A Manual of the Insects of the Plains* (London: W. Thacker, 1909), https://archive.org/details/indianinsectlife00maxw/page/n8.

55. Robert Barde, "Prelude to the Plague: Public Health and Politics at America's Pacific Gateway, 1899," *Journal of the History of Medicine and the Allied Sciences* 58 (2003): 153–86.

56. Francis M. Munson, *Hygiene of Communicable Diseases: A Handbook for Sanitarian, Medical Officers of the Army and Navy and General Practitioners* (New York: P. B. Hoeber, 1920), 515–16.

57. Harold Russell, *The Flea*, vi–vii.

58. Johnson, *Ordering Life*, 296.

59. G. H. E. Hopkins and Miriam Rothschild, *An Illustrated Catalogue of the Rothschild Collection of Fleas in the British Museum (Siphonaptera)* (Oxford: Oxford University Press, 1953), 1–2.

60. "Itching to Know How Fleas Flee? Mystery Solved," NPR, February 10, 2011, https://www.npr.org/templates/transcript/transcript.php?storyId=133602679; "Mystery of How Fleas Jump Resolved," YouTube video, uploaded February 9, 2011, https://www.youtube.com/watch?v=mcnoL1kJ4so.

61. Walter Sullivan, "Miriam Rothschild Talks of Fleas," *New York Times*, April 10, 1984, https://www.nytimes.com/1984/04/10/science/miriam-rothschild-talks-of-fleas.html.

62. "Dame Miriam Rothschild, a Scientist of the Old School, Died January 20th, Aged 96," *The Economist*, February 3, 2005, https://www.economist.com/obituary/2005/02/03/miriam-rothschild.

63. Brendan Lehane, *The Complete Flea: A Light-Hearted Chronicle—Personal and Historical—of One of Man's Oldest Enemies* (New York: Viking Press, 1969), 95.

64. D. A. Humphries, "The Mating Behaviour of the Hen Flea Ceratophyllus Gallinae (Schrank) (Siphonapter: Insecta)," *Animal Behavior* 15, no. 1 (1967): 82–90. doi:10.1016/s0003-3472(67)80016-2.

65. Miriam Rothschild, quoted in Lehane, *The Complete Flea*, 95.

66. George Poinar Jr. and Roberta Poiner, *What Bugged the Dinosaurs? Insects, Disease, and Death in the Cretaceous* (Princeton, NJ: Princeton University Press, 2008), 1–16, 135–39.

67. Asian Scientist Newsroom, "Giant Prehistoric Fleas Found from Mesozoic Era of China," *Asian Scientist*, March 2, 2012, https://www.asianscientist.com/2012/03/in-the-lab/giant-prehistoric-fossil-fleas-found-in-mesozoic-jurassic-cretaceous-period-in-china-2012; Brian Switek, "Super-Sized Fleas Adapted to Feed off Dinosaurs," *Nature*, February 29, 2012, https://www.nature.com/news/super-sized-fleas-adapted-to-feed-off-dinosaurs-1.10135.

68. I wish to thank Claudia Stevens for permission to use passages from her play *Flea*.

69. Marion Wright Edelman, "You Just Have to Be a Flea against Injustice," https://www.goodreads.com/quotes/287480-you-just-need-to-be-a-flea-against-injustice-enough.

70. "American Liberation Front," Petside.com, accessed February 13, 2019, http://animalliberationfront.com/Philosophy/Morality/Biology/RightsOfFliesAndFleas.htm.

71. See, e.g., Chuck Jolley, "Animal Rights Groups a Lot Like Fleas," Feedstuff, January 23, 2015, https://www.feedstuffs.com/story-animal-rights-groups-a-lot-like-fleas-commentary-62-123149.

72. Wayne Covil and Scott Wise, "Woman Jailed after Fleas Kill Her Dog," Channel 6 News Richmond, December 8, 2015, https://wtvr.com/2015/12/08/fleas-kill-dog/.

73. "Flea Diseases in Pets," Pet Basics from Bayer, accessed February 13, 2019, https://www.petbasics.com/dog-education/dog-flea-diseases/.

74. Robert Taber, *War of the Flea: The Classic Study of Guerrilla Warfare* (New York: L. Stuart, 1965; repr., Washington, DC: Potomac Books, 2002), 20.

75. Taber, *War of the Flea*, 49–50.

76. The Japanese effort to use fleas to kill is chronicled in detail in Daniel Barenblatt, *A Plague upon Humanity: The Hidden History of Japan's Biological Warfare Program* (New York: Harper, 2004), and Jeffrey A. Lockwood, *Six-Legged Soldiers: Using Insects as Weapons of War* (Oxford: Oxford University Press, 2009), 95–127.

77. Umezu Yoshijiro, "Operation PX," *Wikipedia*, last edited October 4, 2019, https://en.wikipedia.org/wiki/Operation_PX.

78. Lockwood, *Six-Legged Soldiers*, 172.

79. Lockwood, *Six-Legged Soldiers* 165–70.

80. International Association of Democratic Lawyers, *Report on U.S. Crimes in Korea* (London: International Association of Democratic Lawyers, 1952), http://iadllaw.org/1952/10/iadl-report-u-s-crimes-in-korea-1952/.

81. Quoted in Lockwood, *Six-Legged Soldiers*, 167.

82. Lockwood, *Six-Legged Soldiers*, 169–70.

83. Barenblatt, *Plague upon Humanity*, 227–30.

84. International Association of Democratic Lawyers, *Report on U.S. Crimes in Korea*.

85. The Department of the Army was forced to release the account of "Operation Big Itch" by a Freedom of Information Act request in 2009. It can be read at The Black Vault, accessed February 14, 2019, http://documents.theblackvault.com/documents/biological/bigitch.pdf.

86. "Biological Weapons," United Nations Office for Disarmament Affairs, https://www.un.org/disarmament/wmd/bio.

87. "Chemical and Biological Weapons," International Committee of the Red Cross, August 8, 2013, https://www.icrc.org/en/war-and-law/weapons/chemical-biological-weapons.

88. Lockwood, *Six-Legged Soldiers*, 121. General Umezu was aware of the Japanese effort to develop biological weapons. See Sheldon H. Harris, *Factories of Death: Japanese Biological Warfare, 1932–1945* (New York: Routledge, 2002).

89. Barenblatt, *Plague upon Humanity*, xiii.

90. Miriam Rotkin-Ellman and Gina Solomon, *Poison on Pets II*, NRDC Issue Paper (New York: NRDC, April 2009), https://www.nrdc.org/sites/default/files/poisononpets.pdf.

91. Flea, quoted in Rhian Daly, "Red Hot Chili Peppers' Flea Calls Donald Trump a 'Silly Reality Show Bozo,'" *NME*, February 3, 2016, https://www.nme.com/news/music/red-hot-chili-peppers-49-1205953.

92. Phoebe Waller-Bridge, quoted in Dusty Baxter-Wright, "This is Why the Lead Character in Fleabag Doesn't Have a Name," *Cosmopolitan* (March 14, 2019), https://www.cosmopolitan.com/uk/entertainment/a26819774/why-lead-character-in-fleabag-doesnt-have-name/.

93. Lehane, *The Complete Flea*, 33.

Chapter 8 • *Rodents Gnaw through the Centuries*

1. Hannah More, "Black Giles the Poacher: with some account of a family who had rather live by their wits than their work. Part I," Eighteenth Century Collections

Online, accessed November 17, 2019, https://quod.lib.umich.edu/e/ecco/004773979
.0001.000/1:2?firstpubl1=1700;firstpubl2=1800;rgn=div1;sort=occur;subview=detail
;type=boolean;view=fulltext;q1=rats#hl1.

2. William Shakespeare, *Macbeth*, in *The Arden Shakespeare*, ed. Richard Proudfoot, Ann Thompson, and David Scott Kastan (London: Methuen, 1982), 1.3.8–9.

3. William Shakespeare, *King Lear*, in *The Arden Shakespeare*, 3.4.135–36.

4. This prohibition is highlighted in Mary Fissell, "Imagining Vermin in Early Modern England," *History Workshop Journal* 47 (1999): 9.

5. Wolfgang Miedler, *The Pied Piper: A Handbook* (Westport, CT: Greenwood Press, 2007), 31–65.

6. There is some disagreement among historians of medicine about the role of rats as a disease vector. See chap. 3 for more on the role of fleas and rats in spreading bubonic plague. It has also been suggested recently that the high death toll from plague in the fourteenth century was because of the pneumonic form of the disease rather than the bubonic. See Ben Guarino, "The Classic Explanation for the Black Death Is Wrong, Scientists Say," *Washington Post*, January 16, 2018.

7. Mary Fissell, "Imagining Vermin in Early Modern England," *History Workshop Journal* 47 (1999): 1–23.

8. Edward Topsell, *The Historie of Foure-Footed Beasts* (1607), http://eebo.chad
wyck.com.ezproxy.proxy.library.oregonstate.edu/search/fulltext?ACTION=ByID&ID
=D20000998574280405&SOURCE=var_spell.cfg&DISPLAY=AUTHOR&WARN=N
&FILE=../session/1545266612_22549.

9. Mary Fissell, "Imagining Vermin," 11–15.

10. *Vermont Republican* (Windsor, VT), March 18, 1816; " 'Like Rats Fleeing a Sinking Ship': A History," Merriam-Webster.com, accessed November 10, 2020, https://
www.merriam-webster.com/words-at-play/like-rats-fleeing-a-sinking-ship-history.

11. Oliver Goldsmith, *The Works of Oliver Goldsmith*, vol. 7 (London: J. Johnson, 1806), 167–75.

12. Georges-Louis Lecler, Comte de Buffon, *Natural History: General and Particular*, vol. 1, trans. William Smellie (London: Thomas Kelley, 1856), 417–48.

13. Ellen Airhart, "Rats Have Been in New York City since the 1700s and They're Never Leaving," *Popular Science*, November 30, 2017, www.popsi.com/new-york-city-rats.

14. John Day, *Lawtrickes Or, Who Would Have Thought It* (Blackfriars, 1608).

15. William Rowley, Thomas Dekker, and John Ford, *The Witch of Edmonton* (London, 1658), http://eebo.chadwyck.com.ezproxy.proxy.library.oregonstate.edu/search
/fulltext?SOURCE=var_spell.cfg&ACTION=ByID&ID=D00000126869030000
&WARN=N&SIZE=156&FILE=../session/1546304482_24979&SEARCHSCREEN
=CITATIONS&DISPLAY=AUTHOR. This play was originally written in 1621 but not published until 1658.

16. Anon, *The Wonderful Discoverie of the Vvitchcrafts of Margaret and Phillip Flower, Daughters of Ioan Flower neere Beuer Castle: Executed at Lincolne, March 11. 1618 Who were specially arraigned and condemned before Sir Henry Hobart, and Sir Edward Bromley, iudges of assise, for confessing themselues actors in the destruction of Henry L. Rosse, with their damnable practises against others the children of the Right Honourable Francis Earle of Rutland* (1619), http://eebo.chadwyck.com.ezproxy.proxy
.library.oregonstate.edu/search/full_rec?SOURCE=pgthumbs.cfg&ACTION=ByID&ID

=99838148&FILE=../session/1588361889_26763&SEARCHSCREEN=CITATIONS
&SEARCHCONFIG=var_spell.cfg&DISPLAY=AUTHOR.

17. Anon, *The Wonderful Discoverie*, 9.

18. Anon, *The Wonderful Discoverie*, 11–12.

19. Merry E. Wiesner-Hanks, *Women and Gender in Early Modern Europe*, 3rd ed. (Cambridge: Cambridge University Press, 2008), 254. See the bibliography, 272–75, for an extensive bibliography of the modern historical works on the witch craze.

20. Deborah Willis, *Malevolent Nature: Witch-Hunting and Maternal Power in Early Modern England* (Ithaca, NY: Cornel University Press, 2018). See also Willis's thoughtful discussion on the historiography of the witch craze, 10–25.

21. Robert Darnton has explored the sexual meanings associated with cats and their association with witchcraft in his classic essay "Workers Revolt: The Great Cat Massacre of the Rue Saint-Séverin," in *The Great Cat Massacre and Other Episodes in French Cultural History* (New York: Vintage Books, 1985), 75–106.

22. Topsell, *Historie of Foure-Footed Beasts*, 394.

23. Mark Kurlansky, *Salt: A World History* (New York: Penguin Books, 2002), 5.

24. Thomas Ravenscourt, *Deuteromelia: Or the Seconde Part of Musicks Melodie, or Melodius Musicke of Pleasant Roundelaies; K.H. mirth, or freemens songs. And such delightful catches* (1609), https://search-proquest-com.ezproxy.proxy.library.oregon state.edu/eebo/docview/2248567190/53F140D57EF9433DPQ/1?accountid=13013.

25. Geoffrey Chaucer, "The General Prologue," in *The Canterbury Tales: A Selection*, ed. Donald R. Howard (New York: Signet Classics, 1969). According to Carlo Ginsburg in his classic work *The Cheese and the Worms*, trans. John and Anne Tedeschi (New York: Penguin Books, 1982), "The age-old hostility between peasants and millers had solidified the image of the miller—shrewd, thieving, cheating, destined by definition to the fires of hell. It's a negative stereotype that is widely collaborated in popular traditions, legends, proverbs, fables, and stories" (119).

26. "The Famous Ratketcher, with His Traveles into France, and of His Return to London," English Broadside Ballad Archive, accessed January 7, 2019, https://ebba .english.ucsb.edu/ballad/20214/xml.

27. Robert Boyle, *Some Considerations Touching the Usefulnesse of Experimental Natural Philosophy Propos'd in Familiar Discourses to a Friend by Way of Invitation to Study It* (1663), 219.

28. The role of "witnessing" by reliable observers is explored in Steven Shapin and Simon Schaffer, *Leviathan and the Air-Pump: Hobbes, Boyle, and the Experimental Life* (Princeton, NJ: Princeton University Press, 1985).

29. On Boyle's attitude toward animal suffering, see Anita Guerrini, "The Ethics of Animal Experimentation in Seventeenth Century England," *Journal of the History of Ideas* 50, no. 3 (1989): 391–407.

30. Jean de la Fontaine, "The Two Rats, the Fox, and the Egg," in *The Original Fables of La Fontaine*, trans. Frederick Colin Tilney (1913), https://en.wikisource.org /wiki/The_Original_Fables_of_La_Fontaine,.

31. La Fontaine, "Two Rats, the Fox, and the Egg."

32. E. P. Evans, *The Criminal Prosecution and Capital Punishment of Animals* (London: William Heinemann, 1906), 4, 41.

33. W. W., *The Vermin-Killer, Being a Very Necessary Family Book, Containing Exact*

Rules and Directions for the Artificial Killing and Destroying of All Manner of Vermin (London, 1680), 4–7.

34. W. W., *The Vermin-Killer*, 4–5.

35. Fissell, "Imagining Vermin," 17.

36. Thomas Beard, *The Theatre of Gods Judgements wherein Is Represented the Admirable Justice of God against All Notorious Sinners* (1642), http://eebo.chadwyck .com.ezproxy.proxy.library.oregonstate.edu/search/fulltext?ACTION=ByID&ID =D20000123256710047&SOURCE=var_spell.cfg&DISPLAY=AUTHOR&WARN=N &FILE=../session/1545955819_11076#Hit1.

37. Thomas Beard, *Antichrist the Pope of Rome: Or, the Pope of Rome Is Antichrist Proued in Two Treatises. In the first treatise, 1. By a full and cleere definition of Antichrist . . . In the second treatise, by a description 1. Of his person. 2. Of his kingdome. 3. Of his delusions* (London, 1625), 406.

38. Jan Janszn Orlers, *The triumphs of Nassau: Or, a description and representation of all the victories both by land and sea, granted by God to the noble, high, and mightie lords, the Estates generall of the vnited Netherland Prouinces Vnder the conduct and command of his excellencie, Prince Maurice of Nassau*, trans. W. Shute Gent. (1613), http://eebo.chadwyck.com.ezproxy.proxy.library.oregonstate.edu/search/fulltext ?ACTION=ByID&ID=D00000998496780000&SOURCE=var_spell.cfg&DISPLAY =AUTHOR&WARN=N&FILE=../session/1545948992_702.

39. Sir John Denham, *The Famous Battel of the Catts in the Province of Ulster* (London, 1668).

40. Edmund Chillenden, *The Inhumanity of the King's Prison-Keeper at Oxford* (1643), 4.

41. Anon, *The Mystery of the Good Old Cause Briefly Unfolded* (1660).

42. *Basiliká the Works of King Charles the Martyr: with a collection of declarations, treaties, and other papers concerning the differences betwixt His said Majesty and his two houses of Parliament: with the history of his life: as also of his tryal and martyrdome* (London: Ric. Chiswell, 1687).

43. *Basiliká the Works of King Charles.*

44. Serenus Cressy, "To the Reader," in *Fanaticism Fanatically Imputed to the Catholic Church by Doctor Stillingfleet and the Imputation Refuted and Retorted* (1672), http://eebo.chadwyck.com.ezproxy.proxy.library.oregonstate.edu/search/fulltext ?ACTION=ByID&ID=D00000117810010000&SOURCE=var_spell.cfg&DISPLAY =AUTHOR&WARN=N&FILE=../session/1548109582_4847#Hit1.

45. Edward Stillingfleet, *An Answer to Mr. Cressy's Epistle Apolegetical to a Person of Honour* (1675), 483.

46. Aesop, *Aesop in Select Fables . . . with a dialogue between Bow-steeple dragon and the Exchange grasshopper* (1689), http://eebo.chadwyck.com.ezproxy.proxy.library. oregonstate.edu/search/full_rec?SOURCE=pgimages.cfg&ACTION=ByID&ID =13044041&FILE=../session/1588361215_18762&SEARCHSCREEN=CITATIONS &VID=96878&PAGENO=1&ZOOM=&VIEWPORT=&SEARCHCONFIG=var_spell .cfg&DISPLAY=AUTHOR&HIGHLIGHT_KEYWORD=.

47. See Matthew Risling, "Ants, Polyps, and Hanover Rats: Henry Fielding and Popular Science," *Philological Quarterly* 95, no. 1 (2016): 25–44, on Fielding's attitude toward natural history and the Royal Society.

48. Henry Fielding, *An Attempt towards a Natural History of the Hanover Rat* (London, 1744), 5–10.

49. Fielding, *An Attempt towards a Natural History of the Hanover Rat*, 9.

50. Fielding, *An Attempt towards a Natural History of the Hanover Rat*, 16.

51. Fielding, *Attempt towards a Natural History of the Hanover Rat*, 22–23.

52. Henry Fielding, *The History of Tom Jones, a Foundling* (New York: P. F. Collier & Son, 1917), https://www.bartleby.com/ebook/adobe/301.pdf.

53. Jonathan Swift, "A Letter to Mr. Harding," in *The Prose Works of Jonathan Swift*, vol. 6, *Drapier's Letters*, ed. Temple Scott (London: George Bell and Sons, 1903), http://www.gutenberg.org/files/12784/12784-h/12784-h.htm#link2H_4_0004.

54. Jonathan Swift, *Gulliver's Travels*, ed. Allan Ingram (Peterborough, ON: Broadview Press, 2012), 142.

55. Swift, *Gulliver's Travels*, 154.

56. Jonathan Swift to Henry St. John Viscount Bolingbroke, March 21, 1730, in *The Works of Jonathan Swift*, vol. 1, Thomas Roscoe (London: H. G. Bohn, 1843), 80.

57. Swift, *Gulliver's Travels*, 181.

58. *The Oxford Magazine / By a Society of Gentlemen*, vol. 8 (1772), 225–26.

59. For the description of this cartoon, see M. Dorothy George, *Catalogue and Personal Satires in the British Museum*, vol. 6 (1938), https://www.britishmuseum.org/research/collection_online/collection_object_details.aspx?objectId=1458129&partId=1&searchText=6775&sortBy=objectTitleSort&page=1. There is another similar cartoon by Thomas Rowlandson titled "The Apostate Jack R., the Political Rat Catcher, 1784."

60. John Day, "The Ile of Gulls as Hath Been Often Acted in the Black Fryers, by the Children of the Revels" (1633).

61. Joel F. Harrington, *The Faithful Executioner: Death, Honor and Shame in the Turbulent Sixteenth Century* (New York: Farrar, Straus and Giroux, 2013), 24.

62. Anonymous, "The Famous Ratketcher."

63. Henry Mayhew, "The Rat-Killer," in *London Labour and London Poor*, vol. 3 (London: W. Clowers and Sons, 1861), 3:10–20.

64. Mayhew, *London Labour and London Poor*, 3:10–20.

65. William Makepeace Thackeray, *Vanity Fair* (1848), 305, http://www.gutenberg.org/files/599/599-h/599-h.htm.

66. Mayhew, *London Labour and London Poor*, 3:15.

67. Mayhew, *London Labour and London Poor*, 3:10.

68. On the role of spectacle in the eighteenth and nineteenth centuries, see Simon Schaffer, "Natural Philosophy and Public Spectacle in the Eighteenth Century," *History of Science* 21 (1983): 1–43; and Martin Meisel, *Realizations: Narrative, Pictorial, and Theatrical Arts in Nineteenth-Century England* (Princeton, NJ: Princeton University Press, 1983). On "slippage" between humans and animals, see Barbara M. Benedict, *Curiosity: A Cultural History of Early Modern Inquiry* (Chicago: University of Chicago Press, 2001), 243; and Neil Pemberton, "The Rat-Catcher's Prank: Interspecies Cunningness and Scavenging in Henry Mayhew's London," *Journal of Victorian Culture* 19, no. 4 (2014): 524.

69. Charles Fothergill, *An Essay on the Philosophy, Study and Use of Natural History* (London: White, Cochran, 1813), 139–42.

70. Mayhew, *London Labour and London Poor*, 3:12.

71. Mayhew, *London Labour and London Poor*, 3:20.

72. Mayhew, *London Labour and London Poor*, 3:12.

73. Mayhew, *London Labour and London Poor*, 3:17.

74. Mayhew, "Preface," in *London Labour and London Poor*, 1:1.

75. On this debate, see the excellent introduction to Henry Mayhew, *London Labour and the London Poor: A Selected Edition*, ed. Robert Douglas-Fairhurst (Oxford: Oxford University Press, 2010); Pemberton, "Rat-Catcher's Prank," 521.

76. Mayhew, *London Labour and London Poor*, 3:11.

77. Mayhew, *London Labour and London Poor*, 3:9.

Chapter 9 • The Two Cultures of Rats, 1800–2020

1. Bram Stoker, *Dracula* (1897), http://www.gutenberg.org/files/345/345-h/345-h .htm.

2. "A Secret Tape, a Rightwing Backlash: Is Michael Cohen about to Flip on Trump?," *The Guardian*, July 26, 2018, https://www.theguardian.com/us-news/2018 /jul/26/is-michael-cohen-about-to-flip-on-donald-trump.

3. Dana Milbank, "I Smell a Rat," *Washington Post*, September 24, 2018.

4. Curt P. Richter, "Experiences of a Reluctant Rat-Catcher: The Common Norway Rat—Friend or Enemy," *Proceedings of the American Philosophical Society* 112, no. 6 (1968): 406.

5. John Calhoun, quoted in Wray Herbert, "The (Real) Secret of NIHM," *Science News* 122 (1982): 92–93.

6. Bobby Corrigan, quoted in Emily Atkin, "America Is Infested with Rats and Some of Them Are the Size of Infants," *Mother Jones*, August 25, 2017, https://www.mother jones.com/environment/2017/08/climatedesk-warmer-weather-ratpocalypse/. See also Robert Sullivan, *Rats: Observations on the History and Habitat of the City's Most Unwanted Inhabitants* (New York: Bloomsbury, 2004), 98. Sullivan's best-selling book contains almost everything one might want to know about the history and habits of rats in the past and the present. Bobby Corrigan was one of the experts he consulted.

7. Karen Houppert, "Oh Rats: There Is One Aspect of Baltimore She Can't Get Used To," *Washington Post*, June 19, 2013.

8. On the common themes in Browning and Stoker, see Sam George, "Spirited Away: Dream Work, the Outsider, and the Representation of Transylvania in the Pied Piper and Dracula Myth in Britain and Germany," in *Dracula: An International Perspective*, ed. Marius-Miricea Crispan (New York: Springer, 2017), 69–87.

9. George Orwell, *1984* (New York: Signet Classics, 1950), 286.

10. Sigmund Freud, *Notes upon a Case of Obsessional Neurosis* (1909), https:// archive.org/stream/SigmundFreud/Sigmund%20Freud%20%5B1909%5D%20 Notes%20Upon%20A%20Case%20Of%20Obsessional%20Neurosis%20%28The%20 Rat%20Man%20Case%20History%29%28James%20Strachey%20Translation%20 1955%29_djvu.txt. On Freud and Orwell, see Paul Rozen, "Orwell, Freud, and *1984*," *VQR* 54, no. 4 (1978): https://www.vqronline.org/essay/orwell-freud-and-1984.

11. Oriana Fallaci, quoted in Ian Fisher, "Oriana Fallaci, Writer-Provocateur, Dies at 77," *New York Times*, August 26, 2006, https://www.nytimes.com/2006/09/16/books /16fallaci.html.

12. Erin Steuter and Deborah Wills, *At War with Metaphor: Media, Propaganda and*

Racism (New York: Lexington Books, 2008), xi. Steuter and Wills highlight the symbolic power of language used by politicians and the media to dehumanize opponents. Although their approach is sociological rather than historical, we discuss many of the same words and images.

13. "Boy with Rat Bites Found Dead in a Tenement Flat in Brooklyn," *New York Times*, January 26, 1964, https://www.nytimes.com/1964/01/26/archives/boy-with-rat-bites-found-dead-in-a-tenement-flat-in-brooklyn.html.

14. Lauren Windsor, "GOP Crowd Applauds Calling Immigrants Rats and Roaches," *Huffington Post*, May 10, 2015, https://www.huffingtonpost.com/lauren-windsor/gop-crowd-applauds-callin_b_7253120.html.

15. "America's Rat Race Sponsored by Capitalism," Daily Kos, May 30, 2014, https://www.dailykos.com/stories/2014/5/30/1303035/-Issues-Under-Fire-The-Great-American-Rat-Race-Sponsored-by-Capitalism.

16. I borrow the phrase "rats are rats" from Bruce Alexander, an emeritus psychologist who taught at Simon Fraser University. See Bruce Alexander, "Addiction: The View from the Rat Pack," accessed August 31, 2018, http://www.brucekalexander.com/articles-speeches/rat-park/148-addiction-the-view-from-rat-park.

17. Robert C. O'Brien, *Mrs. Frisby and the Rats of NIMH* (New York: Simon and Schuster, 1971), 137.

18. O'Brien, *Mrs. Frisby and the Rats of NIMH*, 175. See also Edmund Ramsden, "From Rodent Utopia to Urban Hell: Population, Pathology, and the Crowded Rats of NIMH," *Isis* 10 (2011): 659–88.

19. For the role of Johns Hopkins University and the scientists associated with the university, see Sullivan, *Rats*, 15–20, 231–32.

20. John B. Calhoun, "Population Density and Social Pathology," *Scientific American* 206 (1962): 139–49, 139.

21. Calhoun, "Population Density and Social Pathology," 146.

22. Calhoun, "Population Density and Social Pathology," 148.

23. Fredrick Kunkle, "The Researcher Who Loved Rats and Fueled Our Doomsday Fears," *Washington Post*, June 19, 2017.

24. A detailed description and critique of Calhoun's work can be found in Edmund Ramsden and Jon Adams, "Escaping the Laboratory: The Rodent Experiments of John B. Calhoun and Their Cultural Influence," *Journal of Social History* 42 (2009): 761–92, and their commentary on *John B. Calhoun Film 7.1*, "The Falls of 1972: John B. Calhoun and Urban Pessimism," *Circulating Now* (blog), January 11, 2018, https://circulatingnow.nlm.nih.gov/2018/01/11/the-falls-of-1972-john-b-calhoun-and-urban-pessimism/. Calhoun's papers were released by the US National Library of Medicine in 2012.

25. Tom Wolfe, "O Rotten Gotham—Sliding Down the Behavioral Sink," in *The Pump House Gang* (New York: Bantam Books, 1968), 233.

26. Lewis Mumford, *The City in History*, 2nd ed. (New York: Harcourt Brace, 1968), 210.

27. Quoted in Wray Herbert, "The (Real) Secret of NIMH," *Science News* 122 (1982): 92–93.

28. For a detailed discussion of Richter's scientific work, see the Jay Schulkin, Paul Rozin, and Eliot Stellar, "Curt P. Richter," in *Biographical Memoirs*, vol. 65 (Washington,

DC: National Academies Press, 1994), 310–20, https://www.nap.edu/read/4548 /chapter/17.

29. Richter wrote a memoir about his World War II experiences, "Experiences of a Reluctant Rat-Catcher: Norway Rat—Friend or Enemy?," *Proceedings of the American Philosophical Society* 112 (1968): 403–15.

30. Curt P. Richter, "Rats, Man, and the Welfare State," *American Psychologist* 14 (1959): 18–28.

31. Richter, "Rats, Man, and the Welfare State," 27.

32. Morris E. Eson, "Comment," *American Psychologist* 14 (1959): 593–94.

33. Samuel D. Ehrhart, *The Fool Pied Piper*, 1909, photomechanical print, Library of Congress, Washington, DC, https://www.loc.gov/pictures/item/2011647475/. See also Gregory Pappas, "Trump's 'Undesirable' Muslims of Today Were Yesteryear's Greeks: 'Pure American. No Rats, No Greeks,'" Pappas Post, December 9, 2015, http://www .pappaspost.com/todays-undesirable-muslims-were-yesteryears-greeks-pure -american-no-rats-no-greeks/.

34. For a detailed discussion of the use of animal metaphors, including rat and insect metaphors, on society, politics, discourse and behavior, see Steuter and Wills, *At War with Metaphor.*

35. Lawrence Bush, "March 5: Nosferatu," *Jewish Currents*, March 4, 2015, https:// jewishcurrents.org/jewdayo-grid/march-5-nosferatu/.

36. Quoted and translated in "Der eige Jude: The 'Eternal Jew' or the 'Wandering Jew,'" Holocaust Education and Archive Research Team, http://www.holocaust researchproject.org/holoprelude/derewigejude.html.

37. Julius Streicher in *Der Stürmer*, quoted by Richard Webster, June 30, 2003, http://www.richardwebster.net/print/xofratsandmen.htm.

38. Richard Wright, *Native Son* (New York: Harper and Brothers, 1940), 1–8.

39. Wright, *Native Son*, 428.

40. Wright, *Native Son*, 402.

41. "Death in a Long Dark Tunnel!," *Chamber of Chills* 14 (January 1, 1975): https:// comicvine.gamespot.com/chamber-of-chills-14-death-in-a-long-dark-tunnel/4000 -52473/.

42. Joseph Mitchell, "The Rats on the Waterfront," in *Up in the Old Hotel* (New York: Vintage Books, 1992), 490–91.

43. "Spook," *Oxford English Dictionary*, https://www-oed-com.ezproxy.proxy. library.oregonstate.edu/view/Entry/187392?rskey=NVjw60&result=1#eid.

44. Jesse Gray, quoted in Sullivan, *Rats*, 62.

45. L. C. Gardner-Santana, D. E. Norris, C. M. Fornadel, E. R. Hinson, S. L. Klein, and G. E. Glass, "Commensal Ecology, Urban Landscapes, and Their Influence on the Genetic Characteristics of City-Dwelling Norway Rats (*Rattus Norvegicus*)," *Molecular Ecology* 18 (2009): 2766–78.

46. Glenn Ross quoted in Nick Wisniewski, "The Creation of the Ghetto: An Inter- view with Glenn Ross," Independent Reader, accessed October 16, 2018, https:// indyreader.org/content/creation-ghetto-interview-glenn-ross.

47. Karen Houppert, "Oh Rats: There's One Aspect of Baltimore She Can't Get Used To," *Washington Post*, November 14, 2010.

48. Mark Eckenwiler quoted in Rachel Chason, John D. Harden, and Chris Alcantara, "Rat Complaints Are Soaring and D.C. Is Doubling Down on Its Efforts to Kill Them," *Washington Post*, August 28, 2018, https://www.washingtonpost.com /graphics/2018/local/rat-calls/?noredirect=on&utm_term=.45b9c66f5ee0.

49. Sullivan, *Rats*, 34.

50. Joseph McCarthy, quoted in Bruce Mazlish, "Toward a Psychohistorical Inquiry: The Real Richard Nixon," *Journal of Interdisciplinary History* 1, no. 1 (1970): 82.

51. Richard Nixon, quoted in Jonathan Aitkin, *Nixon: A Life* (New York: Simon and Shuster, 2015), 273.

52. Major Douglas, quoted in Howard Palmer, "Politics, Religion and Antisemitism in Alberta, 1880–1950," in *Antisemitism in Canada: History and Interpretation*, ed. Alan Davies (Waterloo, ON: Wilfrid Laurier Press, 2009), 178.

53. Lianne McTavish, "Rats in Alberta," in Janice Wright Cheney, *Cellar: The Work of Janice Wright Cheney* (Nova Scotia: Beaverbrook Art Gallery / Art Gallery of Nova Scotia, 2012), 58.

54. "You Can't Ignore the Rat!" (1950), in McTavish, "Rats in Alberta," 56.

55. Reprinted in McTavish, "Rats in Alberta," 56.

56. "The Only Good Rat Is a Dead Rat" (1975), *Montreal Gazette*, August 21, 2012, http://www.montrealgazette.com/poster+from+1975+encouraging+Albertans+keep +rats+only+good+dead+reads+poster/7124345/story.html.

57. Alexander, "Addiction"; American Civil Liberties Union, *The Dangerous Overuse of Solitary Confinement in the United States* (Washington, DC: American Civil Liberties Union, 2014), https://www.aclu.org/sites/default/files/assets/stop_solitary_briefing _paper_updated_august_2014.pdf.

58. Ramin Skibba, "Solitary Confinement Screws Up the Brains of Prisoners," *Newsweek*, April 18, 2017.

59. See, e.g., John T. Cacioppo and Louise C. Hawley, "Perceived Social Isolation and Cognition," *Trends in Cognitive Sciences* 10 (October 13, 2009): 447–54; and Adriana B. Silva-Gómez, Darío Rojas, Ismael Juárez, and Gonzalo Flores, "Decreased Dendritic Spine Density on Prefrontal Cortical and Hippocampal Pyramidal Neurons in Post Weaning Social Isolation Rats," *Brain Research* 983, no. 1–2 (2003): 128–36.

60. L. Sun et al., "Adolescent Social Isolation Affects Schizophrenia-like Behavior and Astrocyte Biomarkers in the PFC of Adult Rats," *Behavioural Brain Research* 333 (August 30, 2017): 258–66, doi:10.1016/j.bbr.2017.07.011.

61. Heather Knight, "Poop. Needles, Rats. Homeless Camp Pushes SF Neighborhood to the Edge," *San Francisco Chronicle*, June 24, 2018, https://www.sfchronicle .com/news/article/Neighbors-disgusted-over-despair-on-block-hit-13015964.php.

62. "Homeless Poles Living on Barbecued Rats and Alcoholic Hand Wash," *The Guardian*, August 12, 2010, https://www.theguardian.com/uk/2010/aug/12/homeless -poles-rough-sleepers.

63. Sarah Knaton, "Genetically Mutated Rats Could Be Released in Britain to Solve Rodent Problem," *The Telegraph*, December 5, 2017, https://www.telegraph.co.uk /science/2017/12/05/genetically-mutated-rats-could-released-britain-solve-rodent/.

64. Alissa J. Rubin, "Rodents Run Wild in Paris: Blame the European Union," *New York Times*, December 15, 2016, https://www.nytimes.com/2016/12/15/world/europe /paris-rats.html.

65. Catherin Calvet, "Libération," May 2, 2018, https://www.liberation.fr/auteur /3273-catherine-calvet.

66. Josh Jacobs and Matthew Dalton, "In France, Even the Rats Have Rights," *Wall Street Journal*, August 10, 2018, https://www.wsj.com/articles/in-france-even-the-rats -have-rights-1533914241.

67. Banksy quoted in "Banksy Rat Stencils and Graffiti," Web Urbanist, December 28, 2008, https://weburbanist.com/2008/12/28/banksy-graffiti-art/1-banksy-rat-stencils -and-graffiti1/.

68. Beatrix Potter, *The Tale of Samuel Whiskers, or The Roly-Poly Pudding* (1908), https://www.google.com/books/edition/The_Tale_of_Samuel_Whiskers_or_the_Roly /moFku2UPIEEC?hl=en&gbpv=1&dq=tale+of+samuel+whiskers&printsec=frontcover.

69. O'Brien, *Mrs. Frisby and the Rats of NIMH*, 193.

70. Terry Pratchett, *The Amazing Maurice and His Educated Rodents* (New York: HarperCollins, 2001).

71. Pratchett, *Amazing Maurice and His Educated Rats*, 323.

72. J. K. Rowling, *Harry Potter and the Prisoner of Azkaban* (New York: Scholastic, 1999), 366, 375.

73. Lauri Serafin, "Rats Enjoy Sailing, and Other Unexpected Rat Facts," Public Health Insider, October 30, 2015, https://publichealthinsider.com/2015/10/30/rats -enjoy-sailing-and-other-unexpected-rat-facts/.

74. "Ten Top Reasons to Cuddle a Rat," PETA, accessed November 9, 2018, https:// www.peta.org/features/10-rat-facts-befriend-rodent/.

75. Tommanee McKinney, *A Rat's Tale* (Seattle: Amazon Kindle Editions, 2017), https://www.amazon.com/Rats-Tale-Tommanee-McKinney-ebook/dp/B073RX4PW9.

76. David Covell, *Rat and Roach: Friends to the End* (New York: Viking, 2008).

77. T. Coraghessan Boyle, "Thirteen Hundred Rats," *New Yorker*, July 7, 2008, https://www.newyorker.com/magazine/2008/07/07/thirteen-hundred-rats.

Conclusion • The Power of Vermin

1. Donald J. Trump, quoted on CNN, July 14, 2019, https://www.cnn.com/2019/07 /14/politics/donald-trump-tweets-democratic-congresswomen-race-nationalities /index.html.

2. Ben Zimmer, "What Trump Talks about When He Talks about Infestations," Politico, August 6, 2019, https://www.politico.com/magazine/story/2019/07/29/trump -baltimore-infest-tweet-cummings-racist-227485.

3. David A. Graham, "Trump Says Democrats Want to 'Infest' the U. S.," *The Atlantic*, June 17, 2018, https://www.theatlantic.com/politics/archive/2018/06 /trump-immigrants-infest/563159//.

4. William Ian Miller, *The Anatomy of Disgust* (Cambridge, MA: Harvard University Press, 1997), 50.

5. Ambassador Zeljko Bujas, in a letter to *The Independent*, December 19, 1992, https://www.independent.co.uk/voices/letter-bbc-navety-in-glorifying-a-serbian -war-criminal-as-a-national-poet-1564358.html.

6. Quoted in Robert Fisk, "Ariel Sharon," *The Independent*, January 6, 2006, https:// www.independent.co.uk/voices/commentators/fisk/ariel-sharon-by-robert-fisk -6112586.html.

7. "Rot in Hell, Abu Bakr al-Baghdadi," *New York Post,* October 27, 2019, https://nypost.com/2019/10/27/rot-in-hell-abu-bakr-al-baghdadi/.

8. "Huntingdon 'Polish Vermin' Cards Remain a Mystery," BBC News, November 27, 2016, https://www.bbc.com/news/uk-england-cambridgeshire-38107645; Representative Curry Todd, quoted in *Racism Review*, November 14, 2010, http://www.racism review.com/blog/2010/11/14/mexican-immigrants-portrayed-as-multiplying -%E2%80%9Crats%E2%80%9D-white-official/.

9. Ken Cuccinelli, quoted in Nick Wang, "Politics," *Huffington Post*, July 26, 2013, https://www.huffingtonpost.com/2013/07/26/ken-cuccinelli-immigration-rats_n _3658064.html.

10. See, e.g., Long Beach Republicans, "A Bunch of Liberal Vermin Republicans Making a Commotion over Nothing, Let the Man Perform and Honor Him," October 6, 2019, https://www.facebook.com/303799209635608/posts/a-bunch-of-liberal-vermin -feminists-making-a-commotion-over-nothing-let-the-man-/2989464444402391/.

11. "Living in Harmony with House Mice and Rats," PETA, accessed November 12, 2020, https://www.peta.org/issues/wildlife/living-harmony-wildlife/house-mice/; "What about Insects and Other 'Pests'?," PETA, accessed November 12, 2020, https://www.peta.org/about-peta/faq/what-about-insects-and-other-pests/.

12. Troy Farah, "Are Insects Going Extinct? The Debate Obscures the Real Danger They Face," *Discover Magazine*, March 6, 2019, http://blogs.discovermagazine.com /crux/2019/03/06/insect-declines-extinction/#.Xc3b-FdKhPY.

Page references to photographs and illustrations are in *italic* type.